# VALUING NATURE AT THE ENDS OF THE WORLD

SOCIETY AND THE ENVIRONMENT

## SOCIETY AND THE ENVIRONMENT

**SERIES EDITORS: DANA R. FISHER AND EVAN SCHOFER**

The impact of humans on the natural environment is one of the most pressing issues of the twenty-first century. Key topics of concern include mounting natural resource pressures, accelerating environmental degradation, and the rising frequency and intensity of disasters. Governmental and nongovernmental actors have responded to these challenges through increasing environmental action and advocacy, expanding the scope of environmental policy and governance, and encouraging the development of the so-called green economy. Society and the Environment encompasses a range of social science research, aiming to unify perspectives and advance scholarship. Books in the series focus on cutting-edge global issues at the nexus of society and the environment.

*Cities in Action: Organizations, Institutions, and Urban Climate Strategies*, Christof Brandtner

*Saving Ourselves: From Climate Shocks to Climate Action*, Dana R. Fisher

*Reforesting the Earth: The Human Drivers of Forest Conservation, Restoration, and Expansion*, Thomas K. Rudel

*Underwater: Loss, Flood Insurance, and the Moral Economy of Climate Change in the United States*, Rebecca Elliott

*Super Polluters: Tackling the World's Largest Sources of Climate-Disrupting Emissions*, Don Grant, Andrew Jorgenson, and Wesley Longhofer

# VALUING NATURE AT THE ENDS OF THE WORLD

## NATURAL CAPITAL AND CLIMATE DISASTER IN BELIZE

PATRICK M. GALLAGHER

Columbia University Press *New York*

Columbia University Press
*Publishers Since 1893*
New York Chichester, West Sussex
cup.columbia.edu



Library of Congress Cataloging-in-Publication Data

Names: Gallagher, Patrick M. (Professor of anthropology) author
Title: Valuing nature at the ends of the world : natural capital and climate disaster in Belize / Patrick M. Gallagher.
Description: New York : Columbia University Press, [2026] |
Series: Society and the environment |
Includes bibliographical references and index.
Identifiers: LCCN 2026004124 (print) | LCCN 2026004125 (ebook) |
ISBN 9780231215664 hardback | ISBN 9780231215671 trade paperback |
ISBN 9780231566513 epub | ISBN 9780231560917 pdf
Subjects: LCSH: Human ecology—Belize |
Natural resources—Belize | Environmental economics—Belize |
Belize—Environmental conditions | Belize—Social conditions
Classification: LCC GF522.B42 G35 2026 (print) | LCC GF522.B42 (ebook) | DDC 304.2097282—dc23/eng/20260406
LC record available at https://lccn.loc.gov/2026004124
LC ebook record available at https://lccn.loc.gov/2026004125

Printed and bound by CPI Group (UK) Ltd, Croydon, CR0 4YY

Cover design: Noah Arlow
Cover image: Ben Popik

GPSR Authorized Representative: Easy Access System Europe,
Mustamäe tee 50, 10621 Tallinn, Estonia, gpsr.requests@easproject.com

*For Matthew, with love, admiration, and gratitude*

# CONTENTS

# ACKNOWLEDGMENTS

Anthropologists are fond of asserting that people are made relationally. That is, we come into being as ourselves through our relations with others. The same can be said of all that we have the privilege to create. This book is my best effort at wrangling the ideas, insights, and acts of care of others into a coherent narrative about living in a complicated, changing world. And, as such, I owe a great many thanks to all who made and supported me as I have developed this project. I am so grateful to build my life through relations that extend across the United States, Belize, and the world.

Above all else, I am profoundly grateful for my family. Sarah, my wife, supported me through cross-country and international moves, many fieldwork stints, and writing retreats. And she is the mother to the three greatest gifts of my life: Maeve, Emmett, and Callan. My children have always kept me rooted in the concerns that are most fundamental to life: love, care, and presence. They challenge me to see the world with freshness and possibility, and they remind me over and over again of what matters in a world that pulls our attention elsewhere.

My parents, Patrick and Linda, and my sister, Meghan (mostly) willingly cultivated my often-obscure interests and passions. My parents gave me the strong sense that I could aspire toward seemingly big goals if I committed to the daily work of achieving them. You might argue that they went a bit too far there—I believed that an NBA career might be possible into my twenties. But they were never subtle about the fact that any such fanciful goals were achieved on a foundation of hard work, copious good luck, and enduring care for others. I may return to the NBA dreams, or priesthood aspirations, but in the meantime, please enjoy this book as recognition and appreciation of your belief in me.

My grandparents all passed before I began college and might be looking down at this path that I have chosen with a bit of confusion. But I am so grateful for all that they made possible. Their love, care, work, and resilience have made so many worlds possible that they could have never imagined. I'm also remarkably fortunate that with Sarah I gained two new sets of grandparents. I've benefited enormously from the loving presence in my life of Donald and Elizabeth Bailey and Paul and Dorothy White. I have a large family with many aunts and uncles, all of whom have played different roles in shaping me and my thinking. My Uncle Michael is the original anthropologist in our family, always observant of the idiosyncrasies of mundane social life, balancing wonder and derision, like all my closest anthropological friends. I'm saddened that my Uncle Bob passed away before I had the chance to write this book because he is one of the few people who I know would read it and want to talk to me about it. He was a deeply curious and caring man and, perhaps more than anyone else, wanted to know about the things that I studied as anthropologist.

Thank you also to Alicia Ditta, Jane Nadel-Klein, John Pemberton, and David Scott for their roles in introducing me to anthropology.

By a strange, and at the times irritating, set of events, I ended up spending the spring of my junior year at the Biosphere 2 Center in southern Arizona. But it was there that I learned to be an ecologist and to think closely and carefully with the landscape, particularly through the mentorship of Jon Titus. It was, strangely, also where I came to meet Paige West for the first time. She came out to Biosphere 2 from New York to give a lecture, and as she spoke, I began to understand for the first time how anthropology could incorporate my expansive interests. Paige advised my undergraduate dissertation project and brought me to Papua New Guinea for my first experience of anthropological fieldwork. That experience was almost enough, but not quite enough, to scare me away from the discipline entirely. There are few people for whom I owe a greater personal and professional debt than her. She is a friend, mentor, and advocate. Her support has led me to virtually every meaningful opportunity of my adult life. Her thinking and ideas have animated so much of what I understand anthropology, and my place within it, to be. I'm grateful for over two decades of friendship.

I owe a great deal of thanks to professional colleagues in conservation who shaped my approach to anthropological inquiry: Stefan Jackson, Cindy Ehrenclou, David Peifer, Melissa Almendinger, and Melissa Mitchell Thomas.

So many of my early ideas for this project were shaped and cultivated by PhD cohort colleagues at Stanford: Stef Bautista, Hilary Chart, Maria Fernanda Escallón, Mark Gardiner, Helen Human, Elly Power, Ian Simpson, and Anna West. Thank you also to the friendship and wisdom of Maron Greenleaf, Jacob Doherty, Nikhil Anand, Bruce O'Neil, Robert Samet, and Hannah Appel. I am eternally grateful for the patient intellectual guidance and feedback of Sylvia Yanigasako. Sylvia is one of the smartest, most generous yet biting thinkers in anthropology.

A conversation with Sylvia is like a road trip for the mind. Thank you as well to the other thoughtful, influential members of my dysfunctional Stanford family: Lisa Malkki, Paulla Ebron, Doug Bird, Rebecca Bird, James Holland Jones, William Durham, Rachel Tongco, Claudia Engel, Maria Manzanares, Kaila Jimenez, Emily Bishop and Jenn "J. K." Kidwell. Thanks especially to Shelly Coughlan and Ellen Christensen.

Thank you to Andrew Mathews for providing the occasional intellectual refuge at UCSC, for copious ideas for reading, and for introducing me to Micha Rahder, a great academic friend and thinker. I'm deeply appreciative as well for the frequent intellectual support, friendship, and snark offered by Julie Velásquez Runk and Rebecca Witter.

I began writing this book way back at Santa Clara University and I am thankful for the collegial, supportive, and respectful community in the Department of Anthropology and the Department of Environmental Studies and Sciences there, including Michell Marvier, Peter Kareiva, Hari Mix, Mythri Jegathesan, Noreen Golden, and all of the members of the "Drink Tank" community.

I was incredibly fortunate to be invited to join the Department of Anthropology at the University of Texas at San Antonio. There I have found a collegial environment with colleagues committed to questions related to humans and their environment. My colleagues carry out work with diverse epistemological foundations in an atmosphere of shared respect. Thanks especially to Thad Bartlett, Luca Pozzi, Steve Kosiba Michael Cepek, Jill Fleuriet, Robert Hard, Laura Levi, and Deborah Moon. I have been particularly grateful to share a hallway with my friend, Jamon Halvaksz, who is willing to start any conversation with me, major or mundane, with a warm shout across the hall. Thank you for having my back.

All of this support helped prepare me for a meaningful research experience in Belize, a place in which I owe a great deal of gratitude to so many people and communities. I owe special thanks to the Government of Belize, the National Institute of Culture and History, Nigel Encalada, Mito Paz, the Coastal Zone Management Authority and Institute (CZMAI), Roberto Pott, Nadia Bood, Mauricio Mejia, Ellen McRae, Jaume Ruscalleda, Erquel Johnson, Miguel Alamilla, Ben and Joanna Popik, Milio and Kristen at Ak-Bol, Melody Sanchez Wolfe, and the people of San Pedro, among many, many others.

At critical moments, I have benefited from feedback and guidance in several tremendous intellectual communities. As the keynote speaker at the University of Texas at Austin New Directions in Anthropology conference, I enjoyed a supportive environment for thinking through questions of kinship and care that emerged in my work. I was grateful for the invitation to present what became chapter 2 in this book as a speaker at Southwestern University. There, Melissa Johnson, her colleagues, and students offered generous and though-provoking feedback. Mel has been an incredible mentor, and I am deeply indebted to her for so deeply enriching my understanding of the political ecology and environmental history of Belize. Most recently, I had the opportunity to present my thinking on seaweed to the Ecology and Culture seminar at Columbia University. Feedback from the bright, curious community of interdisciplinary scholars greatly sharpened and bolstered my thinking.

Thank you to the anonymous reviewers of the manuscript that became this book. Your critical feedback was generous and formative and made this book substantially stronger. I am deeply appreciative of the time and effort that you invested in my work. Thank you to Eric Schwartz at Columbia University Press for taking on this project. And thank you to Alyssa Napier for stepping in as a supportive editor following Eric's departure.

The research for this project was made possible in part through funding from the Ric Weiland Fellows Program at Stanford University, the Stanford University College of Arts and Sciences, Santa Clara University, and the Vice Provost for Research at the University of Texas at San Antonio.

While writing this book, I lost two of my most important academic mentors and friends, Michelle Bezanson and James (Jim) Ferguson. Michelle welcomed me to Santa Clara, and her entire, shimmering social world, with the warmth of someone you were always supposed to meet. When I have doubts about how to proceed in academic life, I ask myself, "What would Michelle do?" She knew that smart people were plentiful in academia—what made people great colleagues and collaborators was kindness, love, and care. These qualities shaped every decision she made. Few could ever match Michelle, but I do my best to honor her legacy.

As I was finishing the final edits for this book, Jim, my dissertation adviser and warm mentor, passed away in California. Jim had an intelligence of great depth and compassion. He observed the world with precision and gentleness. He conveyed great social complexity with an engaging clarity and simplicity. He was the right kind of adviser for me—engaged but trusting.

When I hurriedly turned in my dissertation draft as I tried to graduate before starting at Santa Clara, I was unsure about how Jim might receive it since I struggled to share much of my draft before then. A couple of weeks later, when we met to discuss the draft, Jim placed the carefully notated, printed-out manuscript bound by an enormous black binder clip down on the table in front of me and said gently with a grin, "You should feel very good about this." If he only knew how good that made me feel. I floated back across the Bay to Oakland that evening. We recently gathered in Palo Alto to celebrate Jim's life and scholarship. His family members, students, colleagues, and friends reiterated the same central theme: For Jim, intelligence and kindness were inseparable attributes. His intelligence made him kinder, and his kindness made him more intelligent. I hope my work as a teacher and scholar, and this book, can be a small extension of his kindness and intelligence in the world.

# VALUING NATURE AT THE ENDS OF THE WORLD

# INTRODUCTION

## The Ends of Nature and Value

*If the world had any ends British Honduras would certainly be one of them. It is not on the way from anywhere to anywhere else. It has no strategic value. It is all but uninhabited. . . . Why then do we bother to keep this strange little fragment of empire?*

—Aldous Huxley, *Beyond the Mexique Bay*

*A cotton-spinning machine is a machine for spinning cotton. Only under certain conditions does it become capital. Torn away from these conditions, it is as little capital as gold is itself money, or sugar is the price of sugar.*

—Karl Marx, *Wages and Capital*

## NATURE AND THE CONSTRUCTION OF COLONIAL VALUE

Britain's colonial project in Belize was, from the very beginning, an incomplete one. The swampy mangrove-dense coastline was ideal for informal economies of furtive pirating but made broader imperial fantasies of a grounded colonial rationality somewhat more difficult to realize. The territory did not give in easily to the material and political relationships of colonialism. Its people, particularly the Maya, actively resisted colonial power, and they were aided by a landscape that seemed to actively defy the rationalization of power to which colonial orders aspired. Belize's shoreline is treacherous and ill-defined. Coral reefs jut to the waterline, creating challenges for seafarers, and, once reached, the land itself presented challenges. It was wet and swampy, covered by dense mangrove forests, the boundary between land and sea never clearly defined.

As a result, Belize has surprisingly little colonial architecture or infrastructure to mark its historical relationship to empire, often much to the disappointment of international travelers. The already limited architectural imprint of colonialism has been further reduced through a history of destructive hurricanes and the slow destruction of humid, salt air. But on a hot, bright late afternoon in 2016, I found myself standing amid an emergent landscape that felt like a late, desperate effort to assert colonial order. I had biked past the main town of San Pedro on the island of Ambergris Caye to the building site of a new "colonial-style" resort hotel. The resort complex was along the rear lagoon of the island—the low, swampy, mangrove-dense area away from the seafront that had largely resisted touristic development over the last thirty years of rapid growth on the island. The former wetland had been covered with a starkly white layer of crushed limestone and coral fragments. Whereas low-lying mangrove forests are typically dark and dense with woody thickets of brown branches and green leaves rooted in shallow water, this landscape was now completely flat, dry, and brightly open. It was as if the hotel developer could only imagine colonizing the mangrove forest by first turning it into a moonscape.

I met Ricardo—a former mainland high school math and physics teacher turned San Pedro deli owner—on the crushed limestone moonscape of the emergent Mahogany Bay Resort on Ambergris Caye, a narrow twenty-five-mile-long island off the northern coast of Belize. Mahogany Bay is not in fact on a bay, though one is currently being constructed from a former coastal mangrove forest. The new buildings are made with Belizean hardwoods that, according to the developer, "emanate a modern take on the country's British colonial past." "Mahogany" and "Bay" are really two words that, more than describing any particular characteristics of this resort, connote a Belize-specific colonial romanticism. The territory was colonized by a loose assemblage of British privateers referred to, and in some circles revered as, "the Baymen." The name highlighted the way they settled furtively into the bays and riverine liminal spaces between valuable lowland forest stands and Spanish trade vessels at sea. These Baymen plundered both resources as part of a generally marginal colonial practice on the fringe of the British Empire. They occupied Belize alongside enslaved African and Indigenous peoples.

This resort, the first international luxury-branded hotel in Belize, ironically was being built by a United States–based investment firm on a part of the island for which there was a long history of colonial *dis*interest. Separated from the languid rear lagoon by a network of smaller unnavigable channels, the space was largely ignored by the Maya, the Mestizos, the British, and Canadian and American tourists. This was a backwater within a country long described as a "colonial backwater." In fact, even now, the luxury resort property abutted, unromantically, the island's sewage treatment plant.

The most authentic part of this colonial reproduction may have been the geography of its labor. The site was in the far south end of Ambergris Caye and down a long driveway off the main road. Like mainland mahogany plantations or island chicle and coconut camps of the past, this site of labor was far from home. As a result, workers building out the enormous new resort ate their lunch on the job instead of making the jaunt into town or going home for the meal, which in Belize is typically the largest of the day and spent as a respite with family.

Tasked with feeding the dislocated workers were Ricardo and his partner Flora. They had been given the job of running a take-out deli, a term in Belize typically used to describe a small restaurant, often informal in structure, that focuses on breakfast and lunch and serves a range of hybridized local cuisine: *garnaches*, *panades*, *salbutes*, and rolled tacos. Garnaches and salbutes were the domain of Flora: "She's the cook,"Ricardo told me. But it was late in the afternoon, and in the back corner of the small but bright white structure, framed by two-by-fours and walled with painted plywood, Flora was already occupied cleaning a whole chicken for tomorrow's stewed breakfast tacos. Ricardo offered to make me a late afternoon snack of salbutes, deep orange corn masa discs fried until puffy and topped with shredded chicken, tomato, onion, and pepper, and he offered me some fresh papaya juice while I waited.

Attuned to the quotidian life of the resort development, Ricardo noted that I did not fit into the typical demographic categories. I—a sweaty, white, thirtysomething American—arrived rosy-cheeked and tired on a muddy and somewhat rusty single-gear bike following a four-mile ride on sandy roads from the north side of the island. A developer would have been better dressed. A potential buyer or curious expat would have at least arrived in a golf cart (the primary mode of transit on the island for those of means) and perhaps have asked for directions to the one functioning full-service establishment in the resort—Rum and Bean, a new high-end coffee shop and rum bar (the temporal boundary lines between coffee and rum consumption were already blurred for island tourists with the resort explicitly offering the true vacation luxury of sanctioned transgression).

And I was not a worker. They were generally either Mennonite timbermen or migrant Spanish workers—not San Pedrano Mestizos—from the mainland (and in many cases neighboring Central American countries), looking for dependable work on this rapidly developing island destination. In a country where racial difference is often historically emplaced, I appeared as a white person out of place.

Ricardo gazed out at me from inside the deli stand as he pressed the corn masa thin for frying, running through this assessment in his head. Then he smiled with a realization: "Are you here to see Johann?"

He was right. Johann, the Belizean-raised child of Austrian immigrants, had partnered with the resort to build an enormous high-end aquaponics greenhouse. He asked if I was interested in Johann's vegetables, primarily a mix of microgreens and arugula (the vegetative foot soldiers of cuisine gentrification). I told him that I heard the greens were quite good, but I wanted to talk to Johann as an anthropologist interested in how people were adjusting to climate change on the island, especially the resurgence of food production there.

"Anthropologist?!" Ricardo replied excitedly.

He then told me about his training in math and physics and his interest in the social sciences and asked me more about my research. I said that my larger interest was in how people value nature and how scientists were coming to describe the parts of nature most valuable to people—like the protection the reef and mangroves provide from hurricanes—as "ecosystem services," a phrase that I feared would sound strange. He nodded and made an affirmative sound that suggested he had more to say.

I took the tightly knotted small plastic bag of papaya juice, now dripping with condensation after transitioning from the chest freezer to the thick, hot Caribbean air. I began to untie the knot at the top. He tapped my hand, saying, "No, no, no—just bite off the bottom corner and squeeze it out from there."

I did that, without great elegance, picking the plastic nub out of my mouth, uncertain where to put it, and settling on my pocket. As I leaned my head back and gently squeezed the slippery bag of juice, Ricardo responded to my description of valuing ecosystems: "It reminds me of this thing I saw [on television] about people selling the moon. They have it divided up [in]to lots and you can buy one. Like people have just given up on the Earth and think you've just got to go buy land somewhere else—there or Mars."

He looked back up at me, waiting for a response, perhaps now concerned that he had led the conversation too far afield.

"It's funny you say that," I answered. "Because it really was research on how to live in outer space that made the ecologists come up with ideas about ecosystem services—it's sort of how they came to value the reefs and mangroves here."

Ricardo interjected, "Yeah it's like that Biosphere 2—I love that place." Now I was taken aback. It was like he already knew the whole strange story I was hoping to tell, and am about to tell, in this book.

"I know!" I replied, perhaps a bit too excitedly. Ricardo was right, Biosphere 2, in the distant high desert of southern Arizona, was a part of this story of how nature came to be understood as a form of capital, a provider of services, and a form of value that conservation scientists hoped could overcome crises of both the environment and economy. In the next ten minutes, I told him an abridged version of the intellectual history of nature, capital, and colonialism that underlies the possibility of imagining intact ecosystems as "natural capital." That ethnographic and historical weaving together of the social life of an emergent ecosystem service science, Cold War ecological research, and the racialized history of territorial dispossession that underlies colonialism in Belize and around the world is the somewhat convoluted story at the heart of this book. It links efforts to imagine space colonization in the Arizona desert to the science of valuing the reefs and mangroves. Ricardo had in that moment unwittingly brought together the narrative threads through which this desolate touristic landscape in waiting—a ravaged wetland filled with the grinded rubble of reefs past on a vulnerable coralline island—revealed the intense precarity of capitalist value production in the colonial landscapes of the Anthropocene.

## NATURE, VALUE, AND COLONIALISM IN BELIZE

The coastal waters of Belize, their flowing, difficult-to-trace forms of connectivity—both social and ecological—and the surprisingly uncertain engagement between forested land and fluid sea prove to be profitable intellectual grounds for following the great social work that goes into producing new ideas about nature and value. As a result of this coincidence of interest in the Belize Barrier Reef and a notable global shift in conservation practice, the reef has become a critical test site for the implementation and evaluation of new conservation practices—in several cases acting as a demonstration site for potentially global new techniques and technologies of environmental management. This has included, among other projects, multiple efforts at valuing the reef system at a variety of scales;[1] projects modeling the flows of ecosystem services (e.g., Patel et al.);[2] educational and advocacy efforts geared toward producing new relationships of responsibility between people and coastal resources that link "ridge to reef" both socially and ecologically;[3] and more recent efforts to quantify "blue carbon" through national carbon stock estimates in mangrove ecosystems.[4]

It has been a remarkable showing of sustained interest in the resources of this small country of approximately 430,000 people, once considered so lacking in value that a Conservative member of the British Parliament in the early 1960s said, "Britain should withdraw from useless places like British Honduras in order to save money."[5] Belize has always felt small and marginal within the larger European colonial project.[6] The country has the land area of Massachusetts with the population of Omaha, Nebraska. It is a part of mainland Central America but is Caribbean-facing, and it is only about seventy miles from the sea to the Guatemalan border. Belize is the only officially English-speaking country in Central America (though a majority of the population now is native Spanish speaking) and Belizean Kriol often dominates casual cross-cultural conversation.

It was this seeming paradox of history and value that initially drew my interest to Belize. I initiated research there with the idea of observing ethnographically as a new form of conservation value and practice "infiltrated" the social and political life of the country. I was interested in the way that an emergent science of valuing nature, based on the idea that ecosystems are a form of natural capital that make ecosystem services valuable to people, came to shape social, ecological, and economic life. My vision was to follow the work of conservation scientists, geographic laborers, and economists as they attempted to translate new universalizing narratives of ecological value into a discourse and practice that is fluid and portable. In some ways, though, I brought with me the very imaginary that I meant to study—that new modes of valuation could be dropped into new social, ecological, and economic spaces and then simply create effects, whether reorienting ideas about value and responsibility or creating the infrastructure for new financial markets. But I found that new models of value were not so much imposed upon this space as, perhaps inadvertently, made through it. Tracing the relations—among people, ideas, and landscapes—that enabled these new ways of understanding environmental value led me to a far more expansive, and at times unwieldy, approach, evidenced by the seemingly disparate ethnographic spaces that are threaded together in this book. My examination of ostensibly "new" forms of value became intertwined with stories about colonial exploitation, natural disasters, and the contours of racial difference.

Part of my initial failure to identify this dynamic of an ongoing racialized, colonial production of value likely speaks to my own positionality as a white man from the United States after years of working in the field of environmental conservation in New Jersey. My own conservation labor had become invested in early efforts to capture environmental value and reorient markets toward protecting environmental resources. Efforts to identify ecosystem services and their value struck me as ripe with potential but also attached to universalizing notions of value that imagined a shared economic logic that shaped all people's interaction with the places in which they dwell and labor. My anthropological senses suggested that producing legible forms of environmental value would be, as social scientists are fond of saying, "more complicated." But I was less attuned to the way the work of valuing nature, socially and technologically, would be deeply entangled with histories of the colonial production of race and conflict. Ideas about human difference and modes of knowing the relational world through Cold War ecological technologies shaped the very conditions for producing value, whether in New Jersey or Belize. Landscapes in which valuable nature is imagined to reside are social productions embedded with histories of labor, life, and exploitation that form not only the landscapes' social meaning but also their very material being. The ethnographic work then became to denaturalize not just the social spaces but the very particular lenses with which I initially encountered them. Part of the sinuous movement in this text is a tracing of my own movement through spaces and ideas as I unsettled my understanding of the historical production of environmental value.

Natural capital and ecosystem services, after all, are not things in the world in any preordained sense. "Nature" becomes "natural capital," to paraphrase Marx, only under certain arrangements of people, knowledge, and material. As the policy manager for a Belizean conservation organization once told me in a moment of frankness, "We can't *actually* show that people are benefiting from ecosystem services. What *is* an ecosystem service anyway?" The contrast between the emergence of explicitly monetary valuations of such services in Belize and the lack of any real effort to make this value circulate as a "thing" via markets has been the source of deep-seated anxiety among some in the country. Like the littoral zone between the tides in which many marine ecosystem services are imagined to exist, these services have a liminal quality that is difficult to possess and engage with. This incomplete economic abstraction emerges from a nascent practice of science that seeks to reimagine the relationship between nature and culture in new and provocative ways.

Despite the country's numerous natural resources and intact forests, researchers were not drawn to Belize by the promise of discovering untapped natural capital and ecosystem services in the same way geologists are drawn to the promise of discovering petroleum reserves merely in need of extraction. Natural capital and ecosystem services must be brought into being. Ecosystem services are attempts to make fungible the complex relational values of nature. This idea of monetary value in intact nature is made socially real through evocative narratives of occurring and potential loss. That is, advocates for the ideas of natural capital and ecosystem services often make this new form of value tractable to general audiences by asking them to imagine the cost of losing the ecosystems that render these services. What if the reef was not there to protect the shoreline from dangerous storms? The production of value and the alarming imagination of imminent disaster go hand in hand. This underlines the strongly affective dimensions of what is often presented as the invasion of a coldly calculative economic rationale into the domain of environmentalism. This affective dimension and the nervous new economies of moral responsibility that it produces are especially important in light of an apparent paradox: Despite the tremendous interest in the economic value of Belize's ecosystems and the anxiety in social science literature around the rapid proliferation of "market nature," no real market in ecosystem services has emerged in Belize (or in many other locations where valuations and markets in ecosystems have been attempted). While there have been efforts to certify conservation areas for carbon credit programs such as REDD, ecosystem services like those valued in the coastal zone have not become free-floating commodities or even entered substantially into the calculative practices of government land-use planning and development.[7]

Yet this interest in valuation, and the preservation of social-material assemblages that make the valuable Belizean landscape, still operate to great effect. Although recent efforts to identify and value coastal marine ecosystem services in Belize have not been the pure form of neoliberal market governance that perhaps only social scientists ever imagined, they have helped to produce a social narrative of nature that dehistoricizes and depoliticizes the current landscape and encourages a relationship with the future shaped by a sense of individualized responsibility for potential total social loss. Ecosystem services create new forms of governance in Belize that operate through what I call an "economy of nerves," not simply a halting form of individualized fear or rational framework for decision making but a socially rendered technology of governance that acts on both the body social and individual bodies in Belize even more powerfully than the presumed calculative logic of capitalism.

I trace this circulation of nerves to the specific experience of the ecosystem service logic interacting with the particular terrain of Belize. Despite naturalizing narratives suggesting that it was somehow passed over by both colonialism and neoliberalism, this is a nature made through the distinctive colonial economy of British Honduras and its independence at the height of the Cold War political order in 1981. Colonial Belize was a relatively small-scale, surgically extractive, enslaved labor–based economy that limited the emergence of private property and agricultural development to maintain a labor pool for work in forest products. This created a landscape that was coproduced with forms of racial and ethnic difference in the country that continue to shape the racialized politics of labor and economy.[8] The "nature" of Belize then is part of the geography made through a deeply uneven social and economic past. Ecosystem service scientists, however, generally fail to think historically and thus often work unintentionally to celebrate and preserve a form of value in the land that represents and reproduces a highly exploitative colonial history.

As part of the effort to both denaturalize and historicize emergent forms of environmental value, I also situate the specificity of this ethnographic work in Belize within a much broader conceptual frame: the Anthropocene. I discuss the Anthropocene concept in the most depth in chapters 5 and 6 as I consider the way new configurations of people, nonhuman living things, and the material world encounter each other on novel terms that "disrupt the liberal humanist epistemologies that both individualize and place humans at the centre of world-making processes."[9] This disruption is destabilizing to existing social and political arrangements but also pregnant with the possibility that emerges with new relational worlds.

The Anthropocene is a complicated, and often controversial, conceptual space in both the natural and social sciences. The term was first popularized in 2000 by the Dutch atmospheric scientist Paul Crutzen, who, along with the biologist Eugene Stoermer, formally introduced it in a piece for the International Geosphere-Biosphere Programme (IGCP) newsletter, "Have We Entered the 'Anthropocene,'? "; in 2002, Crutzen elaborated on the concept in the *Nature* article "The Geology of Mankind." In that piece he argued, "Because of these anthropogenic emissions of carbon dioxide, global climate may depart significantly from natural behaviour for many millennia to come," creating a notional model in which people can fundamentally shape "natural" processes yet are also imagined as external to them. Two decades later, geoscientists remain divided on the term's value, and its geologic meaning, yet the idea has infiltrated the vernacular in both the social and natural sciences as a conceptual device for considering the changes in temporal and spatial scale of human influence on the environment.

Although Crutzen is often credited with popularizing the term, his 2002 article traces a historical lineage to his thinking that links back to the Italian geologist (and Catholic priest) Antonio Stoppani's formulation of the "Anthropozoic" era and the Russian scientist V. I. Vernadsky's idea of the "noosphere" in 1926. This body of thought sought to come to terms with the increasing impacts of human actions across temporal and spatial scales. Research programs in countryside biogeography (discussed in chapter 2) and later natural capital and ecosystem services carry with them this longer historical lineage of research that seeks to understand the way that human actions shape environments.

Most productively for anthropology and the social sciences, debates over the merits of the Anthropocene concept have raised critical questions about the diversity of the human experience, the surprisingly tenuous nature of "humanity" as a concept itself, and thus the drawbacks of imaging a global environment shaped by an epic struggle between a united "Anthropos" and the rest of the world. Anthropologists have pointed toward the analytic shortcomings of imagining a single unified human driver and the way this collapses the very processes of social differentiation that have been central to producing the conditions of crisis we know as the Anthropocene.[10]

Sidestepping the debate about the geologic veracity of the Anthropocene, social scientists have begun to use the concept heuristically as a mode of thinking about the complicated new modes of relationality that shape and configure the contemporary human experience and as an analytic tool for denaturalizing hierarchical assumptions about life that emerged through Enlightenment humanism. As Andrew Mathews puts it, "It may be better to think of the Anthropocene not as a historical epoch defined by geologists but as a problem that is pulling anthropologists into new forms of noticing and analysis and into experiments and collaborations beyond anthropology."[11] Amelia

Moore, building upon Bruno Latour, argues that the Anthropocene is "useful as a confession of sorts of the fallacy that various forms of humanity and earth's biogeochemical processes can each be examined in vacuums that do not contain the other."[12] And, since this book works as an anthropology of both the environment and the scientific practices and narratives by which we come to know that environment, the Anthropocene concept marks a critical epistemological shift in how scientists approach the entanglement of humans with global-scale environmental processes. As Nathan Sayre argues, "The key points to draw from the Anthropocene have less to do with when it began than how it affects the underlying assumptions that scientists make about understanding the world."[13]

Like Moore, I'm using the idea of the Anthropocene as an analytic frame or heuristic for rethinking environmental relationality and knowledge production in the contemporary Caribbean. I'm less invested in debates over whether, objectively, this moment is a different geologic epoch or what the exact limits of human agency are in a complex geologic life world and more interested in the concept's capacity to unsettle received ideas and raise questions about the deeply uneven and long duration of historical relations between people, nonhuman life, and the material world we produce through and with one another. I use this heuristic to diagnose the malignancy of the current moment and to identify and pursue spaces of opportunity and reanalysis in both environmental knowledge and practice. As Deborah Bird Rose has noted, living in the "shadows of the Anthropocene" calls forth a moral obligation to "not abandon" life on the edge and to refuse the violence of the West's "hyper-separation" of nature and culture.[14] But the Anthropocene also at times preemptively refuses that separation and prompts us to ask what ethical commitments we have to emergent forms of life, even ones we never intended or desired to encounter. What does it mean to value a nature that is not just unruly but emergent?

## NATURAL CAPITAL AND ECOSYSTEM SERVICES: AN INTRODUCTION

Partly in response to the critiques regarding the uneven social impacts of protected areas or parks-based conservation, recent environmental conservation efforts have focused on the defense of what has come to be termed "ecosystem services" and "natural capital." Ecosystem services have been defined as "the flows of material, energy, and information from natural capital stocks which combine with manufactured and human capital services to produce human welfare."[15] This focus represents a reorientation of the conservation project from producing a flow of protective services from benevolent people to some natural space or object toward a model of a benevolent nature providing a flow of services toward people. This conservation model aims to safeguard the goods and services from nature in a fashion analogous to ensuring the connectivity and stability of the supply chain for a traditional commodity.

There is also a shift from the protection of a bounded space to the protection of an imagined flow of goods and services from some stock of natural capital. Instead of locating regions of particularly high biodiversity or aesthetic or cultural significance, conservation practitioners have developed various technologies to highlight the ecological components that allow for the functioning of such ecosystem services. These services can come in a variety of forms and are generally broken down into four basic varieties: provisioning (food or genetic resources), regulating (flood control, climate modulation, or other similar functions), cultural (benefits that are nonmaterial, such as a sense of place and spiritual well-being), and supporting (the most basic elements of an ecosystem—nutrient cycling, soil formation, or pollination).[16]

At the root of this conservation strategy is a heterodox economic theory termed "ecological economics." This theory, which is different than classical environmental economics, takes economic systems as a subset of ecological systems and emphasizes that capital is rooted in energy. As a result, ecological economists believe in a form of so-called "strong sustainability" that recognizes limits on the substitutability of natural capital (or the raw material of nature). Ecological economics did not emerge as an organized discipline until quite recently (mainly in the 1980s), but many earlier thinkers (e.g., Malthus, Ricardo) applied similar principles in their critique of classical economics and were influential in the intellectual formation of ecological economics as a discipline.

Critical to the theoretical framework of ecological economics is the idea that natural capital is a unique (through human technology) form of wealth necessary for sustaining life. The concept represents an extension of the classical definitions of capital to include the physical material of nature "such as trees, minerals, ecosystems, and the atmosphere."[17] This is, of course, a marked shift from classical economic theory, which considers the physical material of nature to be a factor of production generally referred to as land or raw material. In the view of ecological economics, the physical material of the natural world is understood as a "stock of natural assets that yields a flow of goods and services," the ecosystem services, thereby making nature an active agent in economic processes.[18]

Early utilization of natural capital as a working concept generally used the term to emphasize the productive significance of nature in ensuring physical and economic health and to unsettle dominant notions of value. It was largely a metaphorical usage intended to broaden dialogue on critical environmental issues.[19] The concept "invited the audience to approach the relationship between nature and economy in a new way with familiar economic terms" and marked an "effort to communicate the delusion of economic growth and the essence of environmental sustainability."[20]

More recently, however, its usage has shifted from a somewhat awkward metaphor to a naturalized economic concept, fostering a growing interest in the explicit monetization of the ecosystem services that flow from natural capital.[21] These monetization schemes imagine ecosystem services as discrete products of nature that can be valued and, in many cases, placed into market-based systems of sale and exchange. Payments for ecosystem services (PES) is the most widely recognized and widely studied. PES is often viewed as a poverty alleviation strategy that targets environmental protection in ecologically and economically vulnerable spaces by determining the value of an identified service and then instituting, according to the World Wildlife Fund website in 2009, a "policy, subsidy, or market to capture this value and reward landowners for conserving the source of the ecosystem services." I offer this example to emphasize the extent to which the abstraction and monetization of these services has proliferated and the degree to which the exchange of money and environmental services has come to stand in for actual social and material investment. I will avoid engaging in the very interesting and potentially problematic implications of these projects for the moment to discuss more closely the ways the concepts of natural capital and ecosystem services can in fact come to fetishize the actual objects of conservation interest.

Proponents of ecosystem services speak of natural capital as nature's riches, a bank of resources that are "undergoing rapid degradation and depletion" at the hands of humans.[22] This framing posits nature's riches as something that *comes into* the hands of humans, not something produced by those hands. The image of humans recklessly raiding the bank of some entity that stands outside of them fails to highlight how humans have been active not only in the depletion of these resources but also in the production of those resources through their ongoing interaction with the landscape. While many proponents of ecosystem services are explicitly moving beyond the idea of a nature-human divide and terms such as "pristine" and "wilderness," the language employed, particularly the unidirectional idea of "flow," still suggests that some entity we refer to as "nature" produces services *for* people rather than a more accurate (and within ecological circles fairly widely accepted) notion that humans are involved in an active cyclical relationship of production with nature.

As Maria Ackerman points out, the metaphors that structure the study of nature are more than mere analogies; they are often unequal relationships of comparison. In the case of natural capital, the "metaphor was reduced to refer to nature as productive machinery. This downplays meanings of nature and natural other than those related to production." The choice of "capital" as a basis for the metaphor of a valuable nature created an opening for the privileging of theoretical and methodological tools of measurement from economics over those of ecology. While this may seem almost inevitable now because of the relative disciplinary powers of these fields, the history (discussed in chapter 3) of ecological research in measuring the movement and economy of energy in the Cold War period meant that the initial debates of methodological and theoretical primacy occurred in the context of a more politically powerful and relevant energy-oriented ecology that had a history of joint modeling with economics.[23] The shift toward thinking of natural capital as an analytical reality waiting to be understood leads to the "mistaken presumption that we can analyze a global problem within a partial equilibrium economic framework and reach a new economy project-by-project without major institutional change."[24]

Beyond an assumption of even political power is an assumption that nature and "the market" are equally innocuous ethical spaces. So, for instance, when a researcher at Arizona State University says, "[Natural capital accounting] is similar to investors being able to see the share price of a firm, assess the capital it holds, and then determine whether they can do a better job managing the company," they imagine a nature that is a benevolent source of wealth and market actors who have an obvious interest in its long-term health.[25] Metaphors such as this one do not include predatory "activist" shareholders such as Carl Icahn and the faceless institutional investor holding companies that make up most actual market activity. Seeing "share prices" in nature, particularly if those shares are undervalued, seems just as likely to lead to exploitation as stewardship.

More recently, a growing number of social scientists in anthropology and cultural geography have engaged with the idea of ecosystem services and market-oriented conservation from a critical perspective.[26] Their work is an extension of the broader critique of what is commonly termed "neoliberalism" in various forms of global development.[27] This work has been productively advanced by studies showing the way neoliberal projects of development and the environment are often quite heterogeneous in practice, and the thread of a potentially more progressive politics can be revealed within them through ethnographic engagement.[28]

Building from this critical insight, a number of works have specifically examined the way markets in environmental services behave in a manner that might be considered antithetical to neoliberalism as an ideology.[29] I build upon, and productively diverge from, these important findings to take what Stephen Collier calls a "technical approach" to various neoliberal "moves" or technologies.[30] Rather than assuming an extant "neoliberal project" that then fails or contradicts itself, I examine the social and intellectual history of the various social threads of ecosystem service science as practiced in Belize. In doing so, I examine how ostensibly neoliberal market-oriented conservation tactics always require site-specific relational improvisations that arrange people, organisms, and materials in a manner that makes it possible to render the abstraction of nature into the abstraction of capital, despite their historic opposition.[31]

## THE COLD WAR AND THE VALUE OF SPACESHIP EARTH

The idea that nature, left as is, could be reimagined as a form of productive capital, made valuable through its anticipated scarcity, came into being with the image of a "global environment" during the Cold War.[32] The combined capacity to view Earth from away with the *Earthrise* photo and the emergent sense that this newly envisioned planet could be threatened in its entirety by ultrapowerful nuclear weapons aided the almost simultaneous development of a notion of planetary belonging and planetary existential threat.[33] I argue in broad historical strokes—drawing on an intellectual history that is part ethnography and part history of science—that ecosystem service science came into being through Cold War–era research concerned with both the threat of nuclear disaster (and the consequent need to trace the radioactive effects of the disaster across a global environment) and the related concern that there may be a need to extend human colonization to outer space in the aftermath of such a disaster. The resulting development of an energetic "systems ecology" and its closed mission cousin, "cabin ecology," laid the intellectual groundwork for a functionalist model of nature concerned with this central question: "What does it take to sustain human life and at what cost?" The ultimate material manifestation of the cabin ecology vision was in the Biosphere 2 project that Ricardo mentioned. The 3.1-acre glass enclosure was an experimental laboratory and proof of concept for creating modular human-sustaining biomes in outer space. Despite being the subject of much ridicule, I will argue in chapter 2, as part of a broader history of the "natural capital concept," that the Biosphere 2 project is an important analytic site for understanding the complex convergence of environmental conservation, market capitalism, Cold War militarism, and the embedded coloniality of ecological modeling. But, for now, note that this intellectual foundation is more than a mere set of functionalist tools; it is made in a social context of both overwhelming nervousness and unease, and vast statist interventions, to develop scenarios by which the potential for disaster could be overcome. I argue then that this social nervousness saturates the tools and narratives of functionalist ecology to fundamentally shape the application and reception of ecosystem service narratives of value in Belize and likely elsewhere.

Much to the surprise of Ricardo (and before that, me), Biosphere 2 may have taught us more about the critical resources or ecosystem services of Belize than it did about living on the moon or Mars. Its enormous ocean biome reproduced a Caribbean coral reef ecosystem like that of the Belize Barrier Reef.

And while the Arizona Department of Agriculture regulations required that the substrate (the limestone and sand) be recreated using materials from within the state, the living corals of the Biosphere 2 ocean came from Belize. Torn from their reef during a storm in the late 1980s as the structure was being built, the corals were transported to their unlikely new home—a small ocean beneath a glass dome in the Arizona desert. And in perhaps the most unlikely connection of all, these domed coral reefs would come by the late 1990s to produce the first scientific research on what is now understood to be the greatest threat to Belize's coral reefs—ocean acidification.

As has been well documented, $CO_2$ levels soared within Biosphere 2 during the human habitation experiment, causing the experiment to be quickly written off as a failure.[34] One of the less visible consequences of this atmospheric change was that the ocean water in the dome became highly acidic. Following the experiment, the marine biologist Chris Langdon attempted to reset the chemistry of the ocean by adding baking soda and baking powder. Each time he tried to do so, it briefly balanced the water chemistry, but the water quickly turned acidic again. When Langdon added those ingredients, he was increasing the saturation level of carbonate ion—the basic building block of coral and other marine calcifiers. This caused the coral to grow more rapidly and thus give off more $CO_2$. This relationship between ocean saturation and coral growth rates was a new finding. As a corollary, Langdon expected and found that with lower saturation levels coral would then grow more slowly, too slowly to keep up with the loss resulting from predation.[35]

That scenario—increased atmospheric carbon, greater ocean acidification, and lower ocean saturation—has become the reality of contemporary oceans in the earth's biosphere, including in the Caribbean Sea. The corals of Belize that were transplanted to Arizona therefore unexpectedly revealed rather precisely the nature of their own looming demise. As problematic and incomplete as the simple functional model of Biosphere 2 was, scientists learned new and important information about the intense relational webs that make ecosystems function productively for humans. If the struggles of Biosphere 2 marked an end to the most enthusiastic period of research into closed ecological systems (capsule ecology), then it may also have been a force in ushering in urgent new methods of ecosystem service science. In fact, in making a narrative transition from an account of Biosphere 2 to a general discussion of ecosystem services, I am quite literally taking a page from the book of the noted ecologist Paul Ehrlich. In *The Dominant Animal* (2008), Ehrlich describes the complexity of the relationships that sustain the Earth's biosphere by providing an account of Biosphere 2's "collapse." "Biosphere 2," he says, "serves as a warning that today's demolition of the ecosystems of Biosphere 1 may portend consequences similar to those in Biosphere 2."[36] From this discussion of the project's shortcomings, he leads directly to the following chapter that introduces the concept of ecosystem services, in a sense arguing literally that ecosystem service science is the next chapter following cabin ecology.

While a certain anxiety saturates much of this work, these sometimes radical reframings of the relationship between nature and culture under the rubric of natural capital or ecosystem services have also opened up new modes of possibility in thinking about value, nature, and responsibility in postcolonial spaces. My ethnographically driven historical approach builds on a growing body of literature that considers the diversity of ways by which existing neoliberal environmental projects work in the world and the surprising political projects they take on and enable. While there is much to be weary of in the various iterations of "market nature" or "neoliberal nature," this growing empirical body of more nuanced substantive analyses suggests two things: Conservation scientists approach their work from surprisingly diverse political perspectives and imagine their work in ways that run counter to simplistic narratives of neoliberal nature; and the tools that ecosystem service scientists have developed have come to be used and circulated in ways that are surprising and perhaps reflective of the diverse politics inscribed in their making.[37]

For instance, Elizabeth Shapiro-Garza describes the way payment for ecosystem service programs in Mexico have been "hybridized through multiple sites of articulation and contestation to become a federal subsidy for rural poverty alleviation."[38] Work by Robert Fletcher in Costa Rica and Maron Greenleaf in Brazil suggest similar outcomes.[39] The regularity of these non-neoliberal outcomes of neoliberal projects suggests that the neoliberal analytic has limitations and that a more technical and historical analysis of these projects opens the possibility for understanding their complex and variegated circulation. I argue that the hybrid or plural qualities of these projects do not simply come into being in the messy cultural space of the field locations of payments for ecosystem service projects (and ethnographies of them) but are actually a part of their original production. Plurality, hybridity, and improvisation are the future of market-oriented conservation projects because they are also their past.

From my ongoing fieldwork in Belize, I have come to think of ecosystem services and neoliberal nature not merely as objects and ideologies whose effects cultural anthropology can ethnographically document but as a scientific and social practice that should be understood and followed as a thoroughly anthropological object of inquiry, something that is made through cultural practices and social interactions between people, institutions, and natural and technological objects. It is important to note, though, that this is very much *not* about revealing science to *actually* be cultural or somehow less real. It is to demonstrate the way scientific knowledges—and the ostensibly standardized models of value that they produce—are necessarily, and often to their great benefit, explicitly social objects.

## THE IMAGINATION OF DESTRUCTION AND THE CONSTRUCTION OF VALUE

During my fieldwork in Belize, conservation scientists asked me in numerous different ways to imagine that the reef was destroyed. They would ask me to imagine a powerful hurricane, a mass bleaching event, or the outcome of an increasingly acidic ocean and then ask me to imagine Belize's coast without the Mesoamerican Reef there to protect it. By contemplating a range of disastrous but entirely possible scenarios, they wanted to distill the value of the reef by imagining a world in its absence. The work of understanding the value of natural capital almost always started with a narrative exercise in imagination, a simultaneous conjuring of both disaster and value.

For all the rigorous science that has been mobilized in the name of ecosystem services, it is fundamentally a practice for seeing nature that requires the practitioner to expand the imagination, to envision a variety of possible futures, all haunted by an overarching imagination of apocalypse. It is no wonder then that the group who developed the Biosphere 2 project began as a theatrical improvisational art group named the Theater of All Possibilities. As the media and public became aware of the Biosphereans' unusual pasts, they often lodged criticisms that suggested improvisation and science were conflicting intellectual pursuits. Building on this critique, in the midst of the contentious human experiments under the dome, one news commentator stood outside of the glass building in the desert, peering in perhaps at those Belizean reefs in the ocean biome, and said, "It looks like something out of science fiction, but how much is science and how much is fiction?" To answer the reporter's question, Biosphere 2, like all modeling in functional ecology, is fiction in that, drawing on Michel de Certeau, it is fabricated through a multiplicity of voices. It is an assemblage meant to convey meaning: "It narrates one thing in order to tell something else."[40] Ecosystem service science attempts to narrate disaster in order, ideally, to provide a venue for telling its audience something about how to live.

In his famous essay "Deep Play: Notes on a Balinese Cockfight," Clifford Geertz declares that the cockfight is an "image, fiction, a model, a metaphor . . . a means of expression."[41] He goes on to declare the expression is somehow "disquietful"—that in its playing-with-fire way of performing social death, it evokes loss as real possibility even if the fight itself is only a performance or simulation. Like the cockfight, ecosystem service science is disquieting because it is a model of value that makes real and possible the usually unspoken promise and possibility of loss and death. Nothing new is made with ecosystem service science, yet by representing the reality of human relations to the environment and each other, it has very real effects.

Later that afternoon, as I was leaving Mahogany Bay Resort, I caught up again with Ricardo and Flora as they walked out the long access road from the resort to the main road to town. I stopped and we talked a bit more. As I prepared to leave them, we stood on the causeway of crushed limestone and sand fill, the decomposed, repurposed remnants of reefs come and gone. Ricardo said, "Look out there to the mangroves," gesturing to our right, away from the resort development.

"And now look back this way," he said, gesturing behind and to our left, back down the road toward the resort. "See, if there's one thing we know about humans, they can adapt the environment to their needs." He paused. "If that's a good thing or a bad thing, I don't know."

There was nothing but flattened crushed limestone between us and the resort, which from this angle appeared to emerge from a void. The three of us, sweat dripping from our foreheads in the low-angle, late afternoon sun, looked out across the dusty white depeopled landscape of the former lagoon.

"But it's home—unless we go to the moon," Ricardo laughed.

Belize, not just as a place of value but as a postcolonial nation, remains in a place of liminality. It is not surprising then that a new technology for representing ecological value, one which produces fantastical numbers such as US $500 million annually for the ecosystem service value of the Belize Barrier Reef, produces anxiety.[42] It is because these visions of value emerge from and perform a long intellectual history of conceptualizing loss in a landscape where loss has been historically linked to racialized colonial taking.[43] And this loss is part of a broader history of the Black diaspora in which the urgency of new narratives of the world's ending is always met with the historical fact in Black studies that "the end-of-the-world is a constant theoretical companion to our decolonial, counterpatriarchical, queer liberatory thinking."[44] Imagine, these models continually ask, the death of the material world that sustains our relational lives. A form of value, then, is made through imagining and performing the existential global threat first made conceivable by the atomic age and now again with global climate change. With acidic oceans that lap higher and higher along Belize's low-lying calcite shoreline—the conveniently accumulated remnants of reefs already lost—and diverse subtropical forests threatened by shifting land use and climate-driven territorial disputes, ecosystem service science in contemporary Belize produces value from extant landscapes by imaging the cost of their increasingly conceivable loss. The result, rather than an empowering form of market-driven rationalized decision-making on resource use, has been a broadening economy of nerves, in which residents and policymakers engage cautiously with value that comes into being only when its disappearance can be most viscerally imagined and where the looming possibility of needing to seek refuge elsewhere—the mainland, the United States, even the moon—looms nervously over daily life.

## CHAPTER OUTLINE

Each chapter in this book concerns some aspect of the production of new forms of value in nature through the interrelated symbolic and material processes of the destruction and construction of ecological space under conditions of anthropogenic climate change. The inquiry emerges from an ethnographic examination of an emergent science for valuing natural capital and ecosystem services but follows the relational threads of this work to consider the production of nature and value more broadly in a country profoundly shaped by global climate change.

In chapter 1, I consider Franz Boas's early foray into anthropology and the perspectival nature of human perception to describe the contingent social, ecological, and historical formations by which a "valuable" nature comes into being in Belize. I describe a tourist at a seaside bar on Ambergris Caye discussing her plan to bottle the various blues of the Caribbean Sea and take them home to enjoy back in her suburban Dallas office. Her misunderstanding of the making of these aqueous colors compelled me to consider the rather complex material, the social and biological factors, that are inherent to human perception and value in nature.

Reminded of Boas's doctoral dissertation—in physics—on the perspectival perception of the color blue, I describe the situated relationship between perception and value, and from that the limitations to the circulation and fungibility of ecological value made in the material and social specificity of the Mesoamerican Reef. The chapter examines that specificity by tracing the circulation of people and resources between the increasingly wealthy and historically Mestizo tourist center of Ambergris Caye and the economically struggling mainland former capital of historically Creole Belize City. In doing so, I begin to trace the relationship between the production of racial difference and the making of value in the Belizean landscape and seascape.

I turn in chapter 2 to describe the complex and surprising social history of natural capital and the underlying ecosystem service science through which it works. While a great deal of work in the social sciences has marked the idea of ecosystem services as "neoliberal," this social history of the concept links its emergence to the development of a functionalist biological systems science for imagining the utter destruction of life on Earth under nuclear annihilation in the Cold War period and subsequent use of that systems science to envision the conditions for space colonization.

The chapter returns ethnographically to consider the Biosphere 2 complex in southern Arizona in greater depth. It was a radical large-scale experiment in the 1980s to develop a prototype closed ecological system for the potential colonization of Mars. From this much maligned but spectacular experiment in systems ecology, I trace the historical and intellectual linkages between these bodies of research to highlight the complex intellectual inheritance of ecosystem service science. I argue that this inheritance may provide an empirical framework for imagining a more plural and inclusive practice for valuing and conceptualizing the relationships between nature and society.

In chapter 3, I take one specific ecosystem service value of interest in Belize, that of landed finfish (the value of large fillet fish caught and sold in Belizean markets), to examine the social gaps and silences in the ecosystem service model of value. I follow along with local small-scale fishermen to understand their role in the circulation of fish from sea to market. This ethnographic tracing reveals the complex and conflicted social relations that shape the social and economic topography of Belizean fisheries along racialized lines. Fish are indeed valuable commodities in coastal Belize, but the models used by ecosystem service scientists often disregard the nuanced narratives of race, ethnicity, labor, and the spatial politics of difference in the postcolonial nation. I contend that producing narratives of sustained ecological flow and function is often dependent on reinforcing narratives of social exclusion and dysfunction that reproduce the coloniality of the Belizean landscape.

My attention shifts in chapter 4 to the technical geographic labor of ecosystem service science. While the emergence of a science for valuing nature is often imagined as the invasion of a blandly technocratic and calculative logic into the realm of nature, I argue here that the embodied labor of mapping ecosystem services in Belize is often deeply laden with affect and emotion. The chapter focuses ethnographically on the work of Geographic Information Systems (GIS) analysts in southern Belize as they attempt to "ground truth" the remotely sensed data about land cover that is used to inform ecosystem service models. I find that GIS analysts often seek out the opportunity to engage in on-the-ground verification of data and in doing so create narrations of landscape and meaning that shape the way they create and interpret ecosystem service models.

In chapter 5, I examine the way tensions in coastal Belize between destruction and production, value and its loss, manifest themselves in evocative ways in debates around the building of a seemingly very mundane object: the seawall. Through a political ecology of the seawall, I account for the ways new narratives of nature and value in coastal Belize have functioned to rework boundaries between nature and culture, water and sea, and economy and society. I describe the productively destructive nature of the littoral zone and the way this zone of meeting between land and sea undermines nearly all attempts to make it more regular, predictable, and valuable through the engineering of seawalls. I maintain that the building of seawalls, and the response to their persistent failure, provides a sort of theory machine for understanding the peculiarly partial, enduringly incomplete abstraction of value that emerges through the ongoing work of what we understand to be market-based conservation.

Finally, in the postscript, I investigate the arrival of an unexpected disruption to the flow of ecological value in coastal Belize: a slimy, putrid, greenish-brown seaweed called sargassum. While sargassum has long circulated in the waters of the Caribbean Sea, it has washed ashore in recent years in enormous volumes that cover beaches, clog nearshore waters, and disrupt fish nurseries and seagrass beds. I explore the way local attempts to theorize the origins of this episodic new disaster (from the *Deepwater Horizon* oil spill to Amazon basin soybeans to African dust storms) reveal the way different actors in Belize understand climate change and the relationship between their local experience of change and larger global forces of power.

The chapter moves ethnographically from these varying narratives of arrival to the work of cleaning beaches in the face of repeated inundations. I show that while government and tourism officials understandably describe the seaweed as an existential threat, some more marginal members of coastal Belizean society see their capacity to clean and circulate seaweed as an opportunity for economic and social gain and as a forced redistribution of political power. In the chapter, I ask how it might change our understanding of, and response to, seaweed inundation if we think of sargassum as a peculiar new form of Anthropocene commons. I argue that these Anthropocene commons are not necessarily formed around objects that are broadly desired as valuable but that chaotic new ecologies can produce unexpected relational possibilities for commoning that counter narrow visions of climate futures.

Taken collectively, the chapters present a threaded relational argument for considering the way colonial landscapes are figured to render capitalist value by virtue of their vulnerability, or anticipated demise, under climate change. I draw ethnographic attention to the historicity of all forms of capital, which even in their most natural form emerge as the outcome of structured social and economic relationships of power. In a Belizean landscape still saturated with an enduring coloniality, natural capital doesn't simply appear; it is rendered from a landscape with a long history of racialized dispossession and exploitation. Valuable nature here, as always, is a product of labor and relations. Through these chapters, I invite the reader to explore and feel the way new forms of capitalist value are made through and haunted by the histories of colonialism, global conflict, and climate change, as well as the historically linked mode of anxious anticipation and regulation that they bring into being. The somewhat nonlinear, threaded nature of this argument, and these chapters, is intended to convey the possibility of other means of relationally constituting this story and thus of imagining value. Part of what drives the feeling of anxiety in this recurring global crisis remains an underlying sense of possibility, the capacity to imagine a future otherwise.

# 1

# THE ONTOLOGY OF BLUE

The Belizean flag, adopted after independence in 1981, is the world's only flag that features humans. It symbolizes (always in ways that escape the intentions of its creators) a particular historical relationship between the country's people, the nonhuman living landscape on which they worked, and the colonial economy that enjoined their histories. The shield at the center of the blue flag depicts two men, one dark skinned and presumably Creole and the other lighter skinned and likely Mestizo. Descriptions of the shield refer to these men as logwood cutters, though they are standing beneath the foliated branches of a large mahogany tree. The flag thereby references the two dominant wood products industries that fueled colonial interest in the territory that would first become British Honduras. Mahogany was valued for its hard, richly colored wood, and logwood was desired for the blue and indigo dyes that could be produced from its heartwood. Together, these two trees drove the colonial economy.

The two men are holding a squaring ax and paddle, respectively, the tools required to cut and then transport the trees from the interior via river out to sea and eventually to Britain. The motto on the scroll just beneath the image of the log cutters reads

"*Sub umbra floreo*"—"Under the shade I flourish." The shield is an awkward representation of the contradictory relationships—colonized and colonizer, nature and culture—enshrined in colonial Belize. The men are workers in an industry that typically exploited enslaved or indentured laborers who were forced to live in remote, muddy, mosquito- and disease-ridden camps. On the flag, they are literally holding the instruments of their own demise. They are poised to cut down the very trees that provide the shade under which they flourish in order to produce products and wealth that they will likely never enjoy. The potential value of this place as a colony and these people as laborers was bound up in the destruction of the very nature imagined as enabling human flourishing. As Malcom Ferdinand has pointed out, the making of value in the Caribbean was always premised simultaneously on a racialized devaluing of bodies, plants, and other animals.[1] This contradiction was literally inscribed in the founding imagery of the nation.

## BOTTLING BLUE

*Sub umbra floreo* perhaps sounds less ironic to the nearly half a million tourists who now visit Belize annually to enjoy the nature that persisted despite and through the specific process of colonial extraction. An impromptu conversation with an American tourist as we sat under the shade of yet another arborous remnant of the extractive colonial economy, the coconut palm, piqued my interest in the contingent ontology of blue and how the social and material histories by which it comes into being reveal a larger story about nature, race, and value in Belize.

On a muggy but brightly sunny July afternoon during my initial fieldwork, I was writing up field notes at the bar of a small yoga resort on a palm-shaded stretch of beach overlooking the Caribbean Sea. Slightly elevated above the water level and positioned at the south end of a subtle point of coralline land that poked at the fringing barrier reef just offshore, it was a remarkable site to enjoy the incredible spectrum of blue that emanates from the waters in the coastal cayes of Belize. Leisure and coloniality are always tightly intermingled in the Caribbean, and here I sat under the shade of non-native palms trees that were the remnants of vast cocal plantations that lined Ambergris Caye and other offshore islands beginning in the late eighteenth century.[2] At the bar there were a few white American tourists, a few Belizean Mestizo and Creole locals, and the primarily Creole members of the restaurant staff.

The intensity and diversity of this spectrum of blues in the Belizean nearshore waters is one of the country's most noted aesthetic qualities. More than mere visual delight, the different perceived colors of water form the basis for a spatial nomenclature that gives some sense of definition, and value, to the vastness that is the sea. What fisherfolk refer to as "the blue" is the darker shade of blue generated by the deeper water outside of the reef, where depths quickly drop from wadable to several hundred feet. Inside the reef is the light blue of "the reef lagoon" (though often now I'll just hear it called the reef, as in "fishing the reef"), and this area is punctuated by oblong darker spaces of blue that can even, depending on the choppiness (and hence sediment level) of the sea, appear a deep green.[3] These areas are referred to as "holes." Despite the dead-end name, they often prove to be profitable foraging grounds for the highly valuable spiny lobster (*Panulirus argus*), which lurks in the cavities along the sandy or muddy walls of these seafloor depressions.

As I sat there gazing out over the sea, pleased with my choice of field site, I was confronted with a challenge of doing ethnographic writing at a yoga resort: A tourist began talking to me.[4] I was writing in a notebook and occasionally typing on a laptop, so this strikingly upbeat woman from Texas sensed that I might be a somewhat unusual brand of tourist.

She turned toward me, smiled, and said, "So, do you live here?"

"For a little while anyway."

"Oh, so what do you do?"

"I'm an anthropologist, actually." For whatever reason, I always add "actually" to the end of that answer.

"Cool, so you study the ancient Mayans?"

"I do cultural and environmental anthropology. I'm interested in how people talk about nature and why they protect it, especially the reef."

"Oh, really! I have a friend who studies marine biology."

Stifling the desire for a more acerbic response, I answered, "Terrific, that's a really interesting field. Are you enjoying your trip?"

"Yes. I could sit here and look out at the ocean all day long. The shades of blue are just *so* amazing."

"Yeah, it really is beautiful," I agreed.

Then she said it, gesturing toward the sea. "I'm thinking of going out there with bottles to collect the different shades of blue to take home." As I listened to her outline a plan that involved a kayak and a friend, I imagined her lining the small bottles up against the gray carpeted wall of a cubicle back in Texas, perhaps next to a photograph of her holding a tropical drink and smiling with the sea glittering in the background.

I laughed uncomfortably. Did she really believe she could bottle up the blue and bring it home? Again, struggling to avoid a more sarcastic response, I simply smiled politely, and said, "Wouldn't that be beautiful."

But wouldn't it be? Wouldn't it be spectacular to carry that contingent beauty with you to a dull office cubicle or the mantel of a suburban home? But, of course, you can't. You can't collect abstract contingent relations and display their colorful outcome in a bottle. The colors of the sea are not fixed and portable (figure 1.1). They are the brilliant and place-specific expression of intersecting relationships between water, its microbial denizens, substrate geology, air, sun, and our particular seeing eyes.

This first chapter seeks, in a perhaps clichéd anthropological move, to "make strange" a mode of valuing nature that many in conservation governance have come to imagine as natural. As described in the introduction (and elaborated on historically in chapter 2), conservation scientists and policymakers have become increasingly enamored with the idea of economically valuing the functional services that intact ecosystems provide to people. This impulse toward monetizing nature has manifested itself in a variety of ways including payment for ecosystem service schemes, carbon credits, habitat banking, and water markets. In Belize, efforts to identify the value of coastal ecosystems have been oriented around what is frequently termed "the

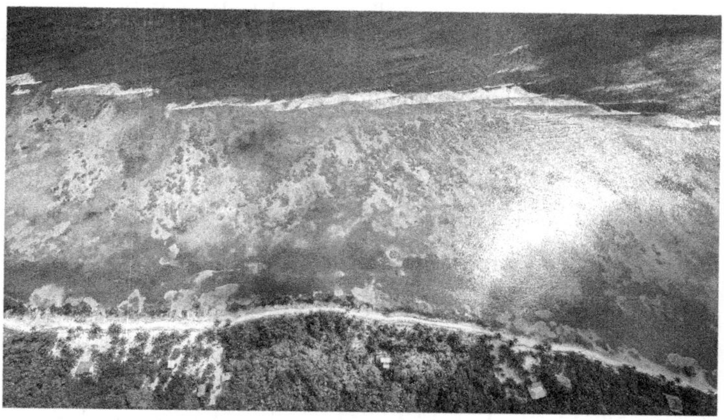

**FIGURE 1.1** The coastline of Ambergris Caye, Belize.

*Source*: Ben Popik

blue economy," a concept that touts the potential for "managing coastal and marine resources in a way that maximizes their economic value to society without jeopardizing ecosystem health."[5] The marine and shoreline environment then becomes a natural asset, a new form of capital, that requires careful management for maximizing ongoing social return. Conservation advocates and government policymakers in Belize have enthusiastically embraced the idea, going so far as to create a new federal Ministry of Blue Economy and Civil Aviation in late 2020.[6] More recently, and remarkably well-aligned with the larger argument of this book, it has been renamed the Ministry of Blue Economy and Disaster Risk Management.

The shift in name gestures toward the close relationship between the production of value and the anticipation of disaster. The blue economy's value becomes newly visible through the punctuated disasters of climate change—its value most visible then at the moment of its imminent loss. But the unsteadiness of the blue economy is not just a fact of climate change; it is bound up in the very notion of the color blue itself—a chromatic concept whose unsteadiness and lack of fixity, its need for ongoing social making—has been central to the making of anthropology, the epistemological frame from which I take on this work, and for the nation of Belize, the space in which I have had the privilege to carry it out. In this opening chapter, I turn to the color blue as an emergent, shifting material and symbolic space through which to theorize on the ontological foundations of both Belize as a nation-state and the nature that undergirds its value as a place in the colonial imagination. I further link this "ontology of blue" to the work of anthropology itself and the specific history by which anthropology became analytically engaged with questions of nature, value, and perception. By linking these instantiations of blue, I make the argument that the bringing into being of Belize was dependent on a certain ontological configuration that was tenuous and contradictory, and that new attempts to locate value in the country's "blue carbon" are inevitably tied up in the longer history of abstraction and circulation that distills the contingent world into the possibility of blue. By tracing the ontological emergence of blue in these spaces, I will trace the intersecting histories of nature, colonialism, and conservation through which Belize's value becomes, in a sense, natural.

Perhaps a yoga resort seems like a strange place for confronting the concept of ontology and the ontological dimensions of nature and value in the colonial world. Ontology is a philosophical frame adopted by anthropology in diverse, even divergent, ways but often focused on notions of fundamental difference.[7] The ontological turn within anthropology broadly describes a growing body of work that examines what is in essence a question of world versus worlds. Do people live in a shared world but with different ways of knowing, interpreting, and making it meaningful—living together with essentially epistemological differences? Or do people actually occupy fundamentally different worlds? That is, are there ontological differences between people?

Considering this framing of fundamental difference, and anthropology's history of engaging with and at times amplifying or exoticizing modes of cultural difference, it is perhaps not surprising that much of the work within the ontological turn comes from spaces that are imagined as classically anthropological field sites: the Amazon basin, Papua New Guinea, and the Indigenous Americas. Anthropological engagement with ontology exclusively in these historically constituted spaces risks becoming a frame for describing radical alterity, for reproducing "the other" in a manner that reproduces uneven colonial interpretive grids in an exploitative manner.[8] There is also some concern that an ontological anthropology, and its discomfort with the nature-culture binary, may in fact miss the "rising purification of those terms as basic political coordinates of contemporary social life," missing the relevance and power of those terms to the very groups of people an ontological turn is imagined to more fully engage.[9]

But Lucas Bessire and David Bond argue ontology also has the potential to "provincialize forms of power within the modern project," in essence, reveal the particular strangeness of Western modes of theorizing what constitutes the world, and thus offer alternatives to the naturalized modes of power that rest on these ontological foundations.[10] Ontology then, as imagined by some within anthropology, becomes a concept that reveals "the multiplicity of forms of existence enacted in concrete practices, where politics becomes the nonskeptical elicitation of this manifold of potential for *how things could be*."[11]

I use ontology here not to wade deeply into recent debates within cultural anthropology concerning the so-called "ontological turn" but rather to call attention to the deeply impactful ways by which groups of people come into being with and through their worlds in profoundly different ways that shape the politics of what seems possible. And I want to call attention to the often rather mundane spaces and moments through which very large ideas about how the world came to be were articulated and contested. It is often through the most mundane aspirations and claims that the most deeply rooted naturalized views are expressed. This is, in some ways, a foundational premise of anthropological ethnography. In this way, I want to peculiarize the mundane goal of the tourist: to unpack, both historically and socially, the accreted modes by which the conventionally Western understanding of nature comes into being and is repeatedly rearticulated and, paradoxically, naturalized as Belize's blue economy becomes an object of value and interest. Questions of ontology in many cases circle back to what Claude Lévi-Strauss famously referred to as the fundamental binary in anthropology—that between nature and culture. That ontological framing, rather than simply unsettling this binary, moves to parochialize it, to contextualize the relatively recent historical and social specificity from which it emerges. It expresses what is now understood as a classically *modernist* ontology that makes the categorical dualism of Western thought appear to predate the very social and historical processes of power that brought them into being and continually reinforce them.[12]

In contrast to the modernist ontology, ethnographic approaches anchored in relational ontologies "emphasize tracing the contextual and contingent paths along which such forms [such as nature/culture and mind/body] come into being, as opposed to populating the categorical spaces of assorted dualist narratives," and thus change "our analytical focus to the ways entities, thought of as processes rather than existents, become entwined."[13]

That said, critiques of the modernist ontology are, paradoxically, often flattening and simplifying. "Modernist science" includes the work of Einstein and quantum theory and has for the past century been critically and reflexively engaged with questions of uncertainty and even the possibility of necessarily postulating the idea of multiple universes to account for "the contradictory bodies of data" that aim to explain and explore the complexity of the universe.[14] The goal of an ontological approach to anthropology, however, isn't to arrive at the "right" one but to engage generously and openly with many. As I critically examine the way market-based approaches to conceptualizing and valuing nature rest firmly on an often unmarked modernist ontology, my aim is to point out the cost of imagining that one particular mode of knowing and being in the world is closer to representing the world in a universal fashion. Such an imagination comes with the loss of possibility and alternative ways of being in a planetary moment when a sense of ontological possibility and plurality might be our greatest foundation for hope.

## NATURE AND CULTURE AT THE END OF THE WORLD

The nature-culture binary is, as anthropologists have long noted, a construction, but it is one with real consequences for the social and material worlds we inhabit.[15] The now orthodox position that the binary is a *cultural* construct, however, often fails to ethnographically interrogate the uneven and highly politicized terms by which that binary is continually made, remade, and contested. The idea of the nature-culture binary as a construct can in narrative practice come to reinforce the binary and do so in unequal ways. John Hartigan argues provocatively that we need to extend our conception of the cultural, to understand it not only as an analytic for the human construction of the nonhuman other but also as an inclusive domain of actors that includes both humans and nonhumans. This approach offers, importantly, the possibility not for simply "getting beyond the human" as some posthumanist and "multispecies" ethnographers call for, but as a means of "pondering what lies between us and them,"[16] and any attempt to understand nature depends deeply on engaging the complexity and multiplicity of the worlds within which humans make themselves.[17]

Market-based conservation projects premised on the idea of ecosystem services are illustrative spaces for considering how the "ambivalent relationship between the human and nonhuman world" comes to be constituted, as well as the differing notions of value (and economy) that are made through these efforts.[18] This project considers how, in these new plans for market-oriented conservation, nature is not merely the object of valuation but the thing *made and remade through* projects of valuation and its spatial representation.

The modern ontology, or what Bruno Latour calls the "modern constitution," is, anthropology suggests, the peculiar one—the mode of constituting and then being in the world that is not particularly enduring and, historically, despite its relatively recent outsized impact, fairly rare. It is a mode of bringing into being a fundamentally colonial worldview.[19] As well, environmental governance is a central political space for reinforcing these ontological claims. Conservation governance, it is said, reinforces the coloniality of the nature-culture ontology through its discursive and material practices.[20]

Stefan Helmreich has argued that while this inherited binary framework in anthropological considerations of nature creates structuring limitations, it has also led to anthropology's unique analytical capacity to press "the nature-culture binary to do a good deal of critical work."[21] If we are to trace the origins of American anthropology to the Boasian school, then we also fundamentally trace the intellectual project of American anthropology to the work of denaturalizing socially produced structures of race and biological difference, of actively reworking the understanding of the relationship between nature and culture. In this sense, anthropological inquiry has long been centered on the dual movement of interrogating the social and cultural modes by which "the natural" comes into being (recognizing these structures as socially real modes of power and authority) *and* developing the interpretive tools for denaturalizing and imagining other ways of constituting our world and our relationships to each other within it.

Particularly within the context of the discipline's commitment to the subaltern, this dual movement acknowledges the fact that the socially and politically disempowered have options that are in fact *structurally* limited. Critical social analysis must recognize the real work that our ideas about nature and the structured relationship between the natural and cultural do in an uneven world.

But this work presents analytical challenges. First, this movement to "denaturalize" at times simply relocates the site of fetishization to culture, making culture an actual domain of all significant phenomena. The reactionary argument that everything we call nature is just culture runs the risk of simply falling back into the very same trap from which we are trying to escape. Recent scholarship in anthropology, science and technology studies (STS), and multispecies ethnography considers methodological and theoretical movements to correct this situation by including theoretical orientations such as the actor-network theory, which allows for a diverse set of "actants" in a "common world," but these analyses also risk flattening that world, a move that anthropology, so attuned to the meaningful topographies of social life, should work to enliven through its empirical perspective.[22] Stories about nature are fundamentally stories about power as manifested in a landscape of diversely agentic actors.

There is a final tension in the anthropological project of deconstructing the nature-culture binary. It is the very dichotomy that brought the discipline into being. The assumed structure of human life makes anthropology an obvious, discrete, and necessary category of inquiry. The continued intellectual move toward breaking down the primacy of the subject-object relationship calls into question our object of study and even more so the methods by which we claim to study it (the object disappears, and the presumed position from which we observe it is made unsteady or decentered).[23]

This interest in nature, culture, ontology, and blue coalesces when considering the origin stories of both anthropology as an academic discipline in the United States and the making of Belize as a colonial nation. The status of blue as a thing, though, has in fact always been uncertain, and this uncertainty over the color, as well as its existence, perception, and universality, were central to the inquiry at the core of North American anthropology. The so-called "father" (even anthropologists can't resist a heteronormative kinship metaphor in relating to our own) of American anthropology, Franz Boas, earned his doctorate in Kiel, Germany, but it was in physics. The dissertation, which was received by his professors without great enthusiasm, was entitled "Contributions to Perceptions of the Color of Water."

The dissertation was based upon "experiments dealing with the absorption of light by different samples of distilled water" that he conducted in Germany.[24] During these experiments, however, he was struck by the challenge of how he as the experimenter came to perceive difference. This pushed him toward considering the intersection of the physical and the psychological, a disciplinary space termed "psychophysics." He claimed that "various differences which traditional psychophysics assumed to be quantitative (e.g., the intensity of light) were in fact *qualitative*."[25]

He would later pursue these questions in the summer of 1883 as part of a journey to Baffin Island in the Canadian Arctic. While the primary aim of his trip was to study the local knowledge of geography, he continued to be drawn to questions regarding the situated nature of perception. Letters to his future wife, Marie Kraockowiser, indicated "a frequent interest in the color of seawater and more generally in problems of perception—for instance, in the situational factors affecting the perception of an iceberg, or the relativity of the perception of temperature." While Stocking argues that the trip may not have been the dramatic anthropological epiphany sometimes later recounted by Boas and his descendants, his time in the Artic did "have a profound and lasting impact."[26]

Boas still identified as a geographer at that point, but the questions that drove Boas, it turned out, were inherently anthropological and, one might argue, ontological questions to the core. Boas's project asked a seemingly simple question about material perception, but through his time in the Arctic, he came to realize that the answer was necessarily a situated one. His project was an early example of a much longer history of anthropological work that has "long called into question the absoluteness of our ontologies."[27] But it stands out here because of its particular interest in how people perceived the color blue in seawater, a materially and semiotically fluid space that unmoored him from the pursuit of ontological certainty.

The mutability of water, Helmreich argues (building on Alexandra Larinia), also served as a model for Boas for the mutability of cultural practice. He maintains that in anthropology, for Boas's descendants, water became an object upon which to meditate, an almost theory machine for the flow and mutability of its central object of interest—culture.[28] Boas's work during this time led him to conclude, as he wrote to his uncle in a revealing letter, "There are domains of our experience in which the concepts of quantity, of measures that can be added or subtracted, like those which I was accustomed to operate, are not applicable."[29] That realization began what Marshall Sahlins calls Boas's "odyssey 'from physics to ethnology.'"[30]

Blue water is a provocatively unstable and challenging construct on which to rest a discipline. In her contemplative essay on the color blue, the literary critic Rebecca Solnit said, "The world is blue at its edges and in its depths. This blue is the light that got lost. . . . This light that does not touch us, does not travel the whole distance, the light that gets lost, gives us . . . the color blue."[31] In the words of the anthropologist Kenneth Little, "Blue is a color of longing," a contingent sensory experience that, though seemingly always just beyond, "emerges momentarily as *something*, some everyday contingent sensibility."[32]

Linguists such as Guy Deutscher have argued that our language comes to shape what we can see and that, in many examples of human civilization, blue is the last color to be named, a puzzle first explored by the British prime minister William Gladstone (in power when Belize, after a long period of tepid occupation, finally became a crown colony in 1871), who noted that the color blue does not occur in the seafaring novel *The Iliad*. For Homer, the sea was "wine-dark." Blue, Little notes, has a history of perception that is uncertain and contingent. And, as with many colors, it came into being as an abstracted object, something that could be essentialized in the form of a dye, through the process of colonialism.[33]

The ability to make the color is likewise central to the origin of Belize as a colonial nation. Logwood (*Haematoxylon campechianum*), prized for its rich dye that could render a range of deep blues and purples, was the first industry in British Honduras. In fact, the industry was thriving long before British Honduras ever became an official colony, and the demise of the market would only prolong Britain's delay in officially recognizing the territory, as Belize's value to the metropole was once again called into question. But early in the colonial period, logwood was tremendously valuable. Its price in the late 1600s was on par with several precious metals, and this created a rare early enthusiasm for British Honduras's prospects as a colony. Sir Thomas Lynch, the governor of Jamaica, declared in 1671 with an enthusiasm that would later wane that the territory "had increased his Majesty's customs and the national commerce more than any of his Majesty's Colonies."

The heart of this valuable tree is a rusty brown. When broken down into small pieces, the wood reacts differentially according to the pH of the agent applied. The logwood chips produce "red colors in an acidic medium and bluer shades in alkaline solutions."[34] The tree became famous in Boston and London for the incredibly valuable liquid blue or indigo dyes that it produced. The logwood, called *eek* by Yucatec Mayans, was long recognized by Indigenous people in the Yucatan for its use as a dye. And the Spanish, who intensely exploited the tree on the other side of the Yucatan in the Bay of Campeche, were not subtle about the nature of their interest, naming the tree *palo de tinte* (dyewood).[35] As trade took off in Belize, in the mid-1600's many groups—the largely British buccaneers leading these early riverside logging expeditions, the growing population of enslaved Africans tasked with the most physical of labor, the merchants in the emergent entrepôt of Belize City—were connected to logwood production. Yet very few of them would ever even see the *blue* of the logwood. People involved in its harvesting and trade likely knew that it had this potential and that there was value in seeing logwood in these terms. But most never saw the blue or the value of this tree-made liquid. Like the blue the woman from Texas imagined holding in a bottle, the blue of the logwood was both not there and structuring what it meant to be there.

Spanish logwood cutting on the Bay of Campeche had started several decades earlier; by the time the practice reached Belize, the tools and techniques for identifying, removing, and valuing the tree were well established. But it wasn't always that way. Consumers in London had to be convinced of logwood's value after initial concerns about the dye's propensity to fade away and its failure to "fix" to cloth. The concerns were so great that an official decree prohibiting its import was passed in 1581 and reiterated in 1619 through the Proclamation for Prevention and Restraint of the Abuses and Inconveniences Occasioned by Dying with Logwood. But the market and perception shifted and red logwood in British Honduras once again became valuable blue dye in England by the mid-1600s.[36] Logwood's value, for both the colony and the larger world market, had to be developed—through the techniques needed to render a commodity from the its raw heartwood and the development of an infrastructure for harvesting the tree, producing and distributing the wood chips, and then exchanging them for money. Logwood was valuable as a dye, but to paraphrase Marx, it becomes a dye "only in certain relationships."

Logwood doesn't merely set the historical stage for the abstraction of value and blue in Belize; it is part of the colonial economic trajectory—in the physical geography of the work and in the use and control of labor that it engendered—which came to materially produce the landscape of contemporary Belize that now makes the country so valuable both on land and at sea. The pattern of settlement established through the history of a specialized wood products economy, first in logwood and later in mahogany, made possible by a relatively small pool of enslaved labor, resulted in a far more limited modification of the forest landscape than in many other colonial contexts and the production of a political and cultural geography uniquely amenable to new notions of value in nature.[37]

## COLONIAL ECOLOGY OF VALUE

In many ways the making of the reef as an object of value and concern has been a remaking of the nation in a way that stands in opposition to its original making. Value has its origins inland and flows and accumulates toward the coast. As a colony, Belize was penetrated from the coast but only glancingly so. It was never a well-surveilled and managed colony. The inland area, for literally two centuries, was not well known or explored by colonial officials. It was in this way a bit of an outlier as a British colony.

The only Anglophone country in Central and South America (though Spanish is now the most common first language), it did not take on the type of plantation economy typical of other Caribbean countries and consequently had a lower economic output, lower population density, and less direct involvement with the slave trade. It also, for the most part, did not utilize the labor, enslaved or otherwise, of Indigenous people as much of Spanish-controlled Central America and parts of the Caribbean commonly did.[38]

Between its initial European settlement sometime in the mid-1600s and through the middle of the nineteenth century, Belize's present-day land was Spain's sovereign territory but settled and economically exploited by Britain. Its economy was dependent almost exclusively on woodcutting: first logwood (used in the European textiles industry) and later mahogany (used primarily for luxury furniture and railroad cars).[39]

Logwood was located primarily in the wet, lowland areas near marshes and rivers; as a result, the British settlement did not penetrate deeply inland. Inland movement was also limited by ongoing territorial conflict with Spain and the presence of understandably hostile Maya communities along the far inland reaches of rivers. A series of agreements in the early 1700s eventually recognized a British right to operate in the territory but maintained Spanish sovereignty and limited the types of activities permitted by the British Baymen. The 1786 Convention of London solidified Britain's log-cutting rights but banned fortifications, formal government, and all agricultural activities, thereby limiting the physical footprint of the British settlers through nearly two centuries of occupation.[40]

From the very beginning then, Belize came to be known in the way that Tim Ingold has described as the spherical orientation.[41] It was slowly penetrated, not surveilled. It was looked into but not looked down upon from above like the colonial gaze from the hill in Cairo that Timothy Mitchell uses to encapsulate the distinctive Orientalist objective position.[42]

The tenuous, evolving nature of its settlement and trade orientation was also influenced by geographic and ecological variables. As the sociologist Nigel Bolland has argued, the "topography and climate of Belize have been very important in the country's social and economic history."[43] In addition to treaties, settlement was limited by the unappealing, swampy coastline, but the system of mangrove islands also provided excellent protection for illicit logwood harvesters who took advantage of the valuable trees' density along the wet banks of the region's sinuous rivers. Belize's nature made it simultaneously unappealing and profitable, an object of colonial interest but largely outside of colonial control, and hence subject to creative and evolving interpretations of its relationships with Spain, Britain, and neighboring territories (especially because of the paucity of early records that would have otherwise more strongly inscribed that history).[44]

When you reach the coastline from the jungle in Belize, the landscape does not simply open to a vast ocean. Fringing the mainland's swampy edge, the nearshore waters are littered with mangrove-covered islands, the tentacled roots of which create an image of almost sci-fi-like active biological resistance to settlement and exploration.

As Joel Wainwright has pointed out, "Until recently, Belizean historiography suggested that all of the Maya people living in (what is today known as) southern Belize were killed before the British arrived—and that the land remained essentially empty until the 1880s."[45] He suggests that this narrative was used in a variety of British colonial settings to justify what he terms a form of "primitive accumulation." Historical accounts of British Honduras often refer to the colony as an "empty wasteland," suggesting that no one would want to inhabit the space as a way of, ironically, justifying the British settlement of the territory.

## ONTOLOGIES OF NATURE

The metaphor that permits scientists to imagine Belize's multicolored coral as a form of natural capital referred to as "blue carbon" (for the ecosystem service of carbon sequestration that it provides) is a contingent and historical formation no less so than that of the red heart of the logwood as a blue dye. Ecologies of value come into being through deeply political epistemic practices. "The relations among and between different forms of life," Tim Choy argues, "are not simply 'out there' to be discovered. . . . They all must be posited and established through scientific (ecological) research."[46] Scientists approach this potential web of relations with deeply rooted views of their world that limit what they can see, as well as what relations they are willing to deem necessary or reasonable. In addition to the relations not simply being "out there," the objects in those relations, and the practices by which we come to distinguish and identify them, must also be posited and established. As Webb Keane has pointed out, while it is somewhat paradoxical to argue with distanced positivist flair that there are different ontologies of the world, the differences by which the world around us comes to be constituted are sometimes so fundamental that they might be understood as entirely different ontologies of nature.[47] Keane argues that for all the challenges of bringing philosophical notions of ontology into ethnographic practice, this challenge might most immediately remind us that ethnography requires the "practical ability to learn how to live with other people, and continues with our taking up the ethical demands that ability entails."[48]

An ethnographic practice that seriously considers not just other ways of knowing but the ways by which a multiplicity of objects can be made real within a common world begins most fundamentally with the work of understanding how people inhabit their world and how they might make, at least temporarily, a home with them. This is, of course, also at the very core of the idea of ecology, which considered through its Greek roots is the study (-ology) of home, house, or family (*oikos*). It is concerned with questions about how organisms, human and otherwise, come to inhabit their world and make it their home. But just as physics, that seemingly universal study of matter and energy, couldn't tell Boas all that he needed to know about how the color blue came into being for the people of Baffin Island, neither can ecology explain all there is to know about how people come to dwell and produce relations of value with nature in

Belize. Value is made in that process of dwelling, and dwelling is in the end central to the idea of ontology as a mode of being in the world.

The practice of environmental conservation around the world increasingly brings its nature into being through philosophical orientations and empirical accounts that draw broadly from both ecology and anthropology. My ethnography then is a deeply entangled engagement in and between these two fields to understand how people come to dwell in a new science and landscape of value in nature.

## CREATIVE DESTRUCTION

The blue waters really were a remarkable sight that day. They brought into strict relief the social geography of the sea. And I was sympathetic to the longing for abstraction described by the woman from Texas, the longing to locate with certainty and make portable and lasting elements of a beautiful but contingent world. But what was behind the somewhat condescending snickers that rattled through the bar that afternoon was a situated knowledge so profound that it could be expressed by simply laughing at the woman who *didn't* know. Abstracting that blue, taking it out of the context—ecological and social—of its being, quite literally destroys it. The blue—the beauty, the value—would no longer exist. You would be left in Dallas holding a glass bottle of water of no apparent color.

But I do not doubt that in that moment of absurdity in which the patrons of the beachside bar laughed as they imagined this woman destroying the blue she sought that she was at the same time bringing it into sharper relief for all of us. We all were forced for a moment to dwell on these striking colors and consider how they can only possibly come into being as such right there—from that slightly raised perspective, with that sunshine, with that dry season water clarity, with our seeing eyes, and with our long history of even coming to know the water and the sky as blue.

This process of valuing natural assemblages through a somewhat convoluted but at times surprisingly productive mode of destruction through abstraction is in fact central to the work of market-oriented conservation approaches concerned with the protection of ecosystem services. The process of valuing nature is what brings that form of nature into being; it doesn't have any preexisting status as such. But conservation scientists intent on valuing nature often engage in a seemingly paradoxical move: To make a valuable nature real, first they imagine destroying it. This initially struck me when a director at the Belize office of a global conservation NGO said, explaining to me the ecosystem service concept, "Imagine there is no reef—what would the cost be to our coastal communities?"

"But where did the reef go?" I asked, only partially joking. I was genuinely interested in where nature must go in order for it to be valued.

"Well, imagine it was destroyed by a hurricane, or damaged by coral bleaching, or rising sea temperatures. All of these are real threats, possible outcomes of a 'business as usual' scenario."[49] By imagining that the reef was not there, destroyed by some nefarious assemblage of nature, she could then begin the process of describing why it was valuable. This, I soon realized, had developed into a standard narrative technique of conservation scientists, one that on reflection aligned in obvious ways with the history of ecological systems thought, rooted as it was in the history of nuclear warfare and the circulation of radiation through the Cold War environment. Ecological models of nature's value were manifestations of a narrative that made nature valuable by imagining first what would happen if we destroyed it. That difference in the value of *other* things (properties, buildings, water, human lives, etc.) between when some element of natural capital was there and when we imagined it was gone is how, for these conservation scientists, ecosystems came to possess value.

Value is therefore made in the space between what is and what might not be. This is a productive and anxious space. Like Heidegger's work, valuing nature is not simply about being but also about not being. Much of people's desire to so intensely interpret the world around them stems from their own limited temporality—they are mortal. What "is" then emerges in part out of the anxiety for humans of what might soon not be. Market-oriented forms of ecosystem valuation are at their core an interpretive grid for the natural world that is mobilized by and brought into relief by the fear of or anticipation of that natural world's disappearance. That disappearance is at once made very real through processes of detailed simulation but will likely never be experienced by the very people that simulate it because of their own mortality. The desire to mark and interpret the value of nature is an at times frenzied, anxious, and apocalyptic narrative art ripe with the anxiety of knowing that both nature and its beholder are not constant, and they emerge in this imagined subject-object relationship from the process of coming to know that.

Nature has no intrinsic value. Like the distinctive color blue that draws us to this marine resource to begin with, it has no set, preexisting ontological status as a thing at all. Yet that kind woman from Texas paid my neighbor—BZ $20—to fetch the different shades of blue from the glistening Caribbean waters, the red heart of logwood was shipped across the open sea to Britain—100 pounds sterling for each ton—to make blue dye in Britain, and scientists now value the reef—$14–$16 per square foot of "vulnerable coast shielded"—for the storm protection that it provides.[50] It is of relatively little ethnographic interest that these ways of seeing, valuing, and exchanging this place are constructs; they rest on forms of value that don't "fix" unyieldingly to the paper on which we write these stories. The aim here is not to merely call attention to the particularity of their narratives or the shifting nature of the landscape through which they produce them. It is to point out that producing a valuable nature takes tremendous work, and in revealing the social labor that goes into producing the narratives, techniques and practices necessary for making nature valuable, a much broader social landscape is revealed in coastal Belize where narratives of nature, society, value, race, ethnicity, and colonialism join together. It is a productively fluid space in which to see how, in our many attempts at valuing nature, we engage in the making and unmaking of whole ways of seeing and being in the world.

# 2

# NEOLIBERAL ECOLOGISTS IN OUTER SPACE

## IF A FAKE RIVER FLOWS INTO AN ECOSYSTEM SERVICES CONFERENCE, DOES IT RENDER SERVICES?

If you looked right toward South Mountain, the setting sun reflected against the windows and glass pool enclosures of Phoenix's encroaching suburban frontier. But if you looked left, the sun interacted more gracefully with the chalky rusted rock of the Arizona desert. The dense spikes of the saguaro (*Carnegiea gigantea*), backlit by the setting sun, created a silken fringing haze for these distinctively "arborescent" cacti. Ocotillo (*Fouquieria splendens*), only recently revealing small bright green leaves from early winter rain, awkwardly launched their dry tentacled branches toward the now purplish dusk sky. It was December 2010, early in my graduate studies, and I had flown from northern California to the Arizona desert for a large gathering of ecosystem service scientists at a resort hotel on the outskirts of Phoenix. The complex was a joint-venture development project of the Gila River Indian Community and located on their tribal land (figure 2.1).

FIGURE 2.1 Aerial view of the Wildhorse Pass Resort outside of Phoenix, Arizona.

*Source*: flickr.com/people/neepster

I enjoyed the view that December evening from a small twelve-passenger ferryboat that was slowly traversing a reproduction of the Gila River. The original Gila River linked the communities of the interior Sonoran Desert to the Colorado River and eventually the Gulf of California, creating a vital linkage that literally brought the arid desert floor to life in collaboration with a complex socio-hydrologic system that developed over thousands of years.[1] The river linked the various O'odham speaking groups that lived in, and continue to live in, present-day southern Arizona, extreme southeastern California, and the Mexican state of Sonora. Today's reproduction of the river, on the other hand, links the back door of the Sheraton Wild Horse Pass Resort and Spa with the Wild Horse Pass Resort and Casino via a meandering two-and-a-half-mile course.

The real river, which still runs consistently blue on a map despite its inconsistent flow, was once so powerful that it was the naming inspiration for a complex society (Akimel O'odham translates to "river people") and was, between 1848 and 1853, the border of two powerful new nations. Today, it is more often a dry sandy ditch occasionally brought to life by the runoff of monsoons, only a fraction of its historic flow ever reaching the Colorado River. Shaded by mesquite, cottonwood, and palo verde trees, the new Gila River was created not simply as a toy river. It attempted to recreate a space in which a whole system of meaning around the riparian landscape of the O'odham people could be narrated to visitors. But in narrating with such deep commitment this obvious simulacrum, it acts to make the enduring absence of the real Gila even more profound.

About half of the posted signs detailed the vegetation of the river and the surrounding desert and described the usage and significance of these plants to the Pima and Maricopa people. This type of interpretive work was tremendously appealing to my boatmates that evening. They had gathered at the resort as part of the ACES (A Community of Ecosystem Services) conference.[2] It is one of the largest scientific conferences in the world focused on ecosystem services. Scientists, programmers, and policymakers gathered for four days to discuss the latest advances in ecosystem service science research, with a particular focus on the computer modeling of ecosystem services and the spatial representation of their value.

The interpretive signage that we encountered essentially narrated what the conference attendees would describe as the ecosystem services of the former Gila River. "Look," one exclaimed, "It's like they made these signs for *us*." In a sense, they did. The river and its signage were an explicit effort to reinscribe, or retain, meaning and value in the crassly gridded landscape of greater Phoenix that now reached out to the reservation (aided by the community's own development arm, the Wild Horse Pass Development Authority).

## THE INTELLECTUAL INHERITANCE OF MARKET NATURE

Conservation scientists have increasingly turned over the last three decades to "market-oriented" techniques to convey the value of "nature" and to promote their desired conservation outcomes. The science of identifying and valuing ecosystem services—"the benefits people obtain from nature"—is at the center of this effort.[3] Both advocates of market nature in the conservation sciences and its critics in the social sciences often describe this shift in practice through a fairly simple origin story about the relatively recent articulation of two seemingly distinct or even adversarial realms—conservation and capitalism.[4] The shared investment in these simplistic narratives of encounter obfuscates more compelling historical narratives that demonstrate the way ecosystem service scientists have inherited a complex and plural intellectual and ideological history that intertwines American market capitalism, Cold War militarism, and the modern environmental conservation movement.[5] These three seemingly distinct, and at times even contradictory, universalizing logics together produced the set of technologies (new kinds of markets, remote sensing and spatial planning technologies, and a relational model of the natural world), the related set of professional actors, and the spaces of intellectual production central to what is understood as the distinctively contemporary realm of market nature and ecosystem service science.

I argue that the elements that are repeatedly understood as destructive internal contradictions in critiques of contemporary ecosystem service science—progressive ecological notions of nature and society, capitalism, and militarized colonialism—are not an emergent contradiction but a complex and often productive intellectual inheritance. This rethreading of the intellectual history of ecosystem services is not an attempt to undermine this emergent science, or those committed to it, but to tell a more plural version of its emergence that creates openings for more diverse social, economic, and ecological futures—ones that embrace the productive historical interplay among these spheres of human action that are frequently policed as distinct or even adversarial.

This chapter describes the way two related realms of Cold War–era ecological inquiry have come to inform so much of the conceptual framework for ecosystem service science: systems ecology, that "robust hybrid of engineering, mathematics, operations research, cybernetics, and ecology"; and cabin ecology and closed systems research.[6] Through an abbreviated history of these spaces of inquiry that traces their material instantiation in the much maligned but quite illuminating history of Arizona's Biosphere 2 project, I aim to link their complex social history to the practice and implementation of an ecosystem service science in contemporary market-oriented environmental conservation programs.

The productive historical intersections between Cold War–era government sponsored research and various "cultures" of science is well-trodden ground in anthropology and cultural geography,[7] science and technology studies,[8] and history.[9] To the extent that I describe that history here, I am not offering something new. My intervention is to link that established history more explicitly to ecosystem service thought to dislodge narratives that imagine market nature as a "disruptive technology" without a history.

In doing so, I demonstrate the complexly variegated intellectual and ideological history of the ecosystem service concept and the possibility for locating a more diverse political future for environmental conservation. Scholars such as Elizabeth Shapiro-Garza, Robert Fletcher, Jan Breitling, and Pamela McElwee account for the regular failure of neoliberal conservation projects by arguing that purer notions of ecosystem services or payments for ecosystem service (PES) schemes become hybridized in practice,[10] but I argue that these projects were hybrid from the beginning. This chapter brings the hybridity into the science's historical and political production rather than arguing that this only happens in the messy field locations of PES's application.

## THE COLD WAR AND THE BIRTH OF ECOLOGY'S GLOBAL VIEW

Ecology emerged as a formalized, and well-funded, discipline through its relationship with the Atomic Energy Commission (AEC) in the 1950s.[11] The techniques and tools of systems ecology—the ways of modeling the relationships of the living world—emerged then through projects of understanding the environmental impacts of highly destructive nuclear weapons and the movement of resultant radiation through the environment.[12] As a corollary, systems ecology developed further through the work of scientists imagining alternative spaces for human evolution in the destruction that might follow, including ambitious closed ecosystem projects like Biosphere 2.[13]

The paradox, of course, is that the same government-funded atomic science that funded and grew a new global ecosystem science during this critical period also made possible the existence of the global environment's end.[14] Many ecologists, whose models were developed through their work (and those of their mentors) in atomic energy, turned their gaze toward outer space to conceive where the evolution of the human species might persist in the wake of nuclear disaster. Through their involvement in space exploration research in a small discipline that came to be termed "cabin ecology," ecologists developed closed system models for sustaining life in space that came to inform the functionalist models of how complex ecosystems actually worked here on Earth.[15] A large part of the intellectual framework for an ecosystem service science emerged from this notion of a closed environment with simple flows of energy aimed at sustaining humans.

By recounting my own movement through the ethnographic field from a large ecosystem service science conference outside of Phoenix to a brief side visit to Biosphere 2 ninety miles away in Oracle, I describe the way these linkages became evident and contextualize them historically before outlining an interpretive framework for evaluating the limitations of current analytical projects concerning the science of market-oriented environmental conservation. I argue that even though ecosystem services are understood as abstractions essential to the making of a market nature, the process of their abstraction occurred within a discipline whose intellectual history is linked to the diverse and often contradictory ideological spaces of central state Cold War planning and the opposing, but necessarily linked, utopian activist ecologies of alternative world-making.

Ecology as an *intellectual discipline* then came into being in critical ways through its work in atomic energy, while ecology as a *political orientation* for remaking global relationships of power emerged through its related work in imagining how human life might persist in the face of disasters wrought by the nuclear. That is to say, the political and ideological roots of ecosystem service science, when viewed historically, are quite diverse and work through systems of meanings and via technologies that reflect diverse and at times entirely contradictory ways of conceiving of and acting in the world.

## DEFINING ECOSYSTEM SERVICES

Conservationists have defined ecosystem services as "the flows of material, energy, and information from natural capital stocks which combine with manufactured and human capital services to produce human welfare."[16] A focus on the protection of ecosystem services, and the natural capital from which they flow, represents a reorientation of the conservation project from producing a flow of protective services from a benevolent humanity to some space or object understood as the "natural" toward a model of a benevolent nature providing a flow of services to humanity. This reorientation began "in the late 1960s and early 1970s [as] a wave of publications by mostly natural scientists was produced which addressed the notion of the usefulness of nature for society, other than an object of ethical concern."[17] It was during the late 1970s and early 1980s that natural scientists came to feel that the most effective way of talking about the usefulness of nature to society was to frame that use in explicitly economic terms. This body of thought congealed around the term "ecosystem services," thanks in large part to the influential piece by Paul and Anne Ehrlich that coined the term.[18] Notably, the book that was credited with creating the concept and injecting it into wider usage was ominously titled *Extinction: The Causes and Consequences of the Disappearance of Species*.[19]

The ecosystem services conservation model focuses on protecting this flow of goods and services from nature in a fashion analogous to ensuring the connectivity and stability of the supply chain for a traditional commodity. In this model, ecosystems (or particular functional relationships within ecosystems) become the form of (natural) capital that produces economically valuable goods and services.[20] All those concerned with nature become both the capitalist and the consumer with a vested interest in investing and maintaining the productive capital, that is, the ecosystems.

The parallel between this linear understanding of ecological production and industrial supply chains is not coincidental. The ecosystems services project involves a very intentional shift from a focus on the protection of a bounded space to the protection of an imagined flow of goods and services from some stock of what has been termed "natural capital." It is part of a larger effort to reimagine the role of nature in capitalist production. This effort moves the locus of conservation work from identifying spaces of particularly high biodiversity, or aesthetic or cultural significance, toward developing various technologies that seek to highlight the ecological components that allow for the functioning of ecosystem services—the abstracted, objectified goods that are produced by intact stocks of natural capital.

Ecologists and economists generally divide ecosystem services into four basic varieties: provisioning (food or genetic resources), regulating (flood control, climate modulation, or other similar functions), cultural (benefits that are nonmaterial such as a sense of place and spiritual well-being), and finally supporting (the most basic elements of an ecosystem—nutrient cycling, soil formation, or pollination).[21] They then seek to spatially represent the provisioning of these abstracted services using GIS technologies and specially designed computer programs for scenario-based conservation planning. The practical politics of these representational practices—what to map, how to map it, how to imagine the end user, the degree toward which the product would structure the decisions of the end user, and the degree of complexity needed or desired by the end user (frequently talked about through an equally profound abstraction, that of the "decision maker")—were the central topics of discussion at the ACES meeting at the Sheraton Wild Horse Pass Resort and Spa that December. Approximately five hundred people had gathered there to consider methods for identifying, valuing and protecting these ecosystem services.

## DAYTRIP TO BIOSPHERE 2

*The finest example of what the human species is capable of inflicting upon itself is Biosphere 2.*

—Jean Baudrillard, *The Illusion of the End*

Beyond mere conference ethnography and sunset boat rides on a reproduction of the Gila River to the casino floor, part of my excitement in traveling to the ACES conference was the opportunity to rent a car and visit Biosphere 2 with a colleague from the conference. It was the somewhat odd place that I had chosen, in the early 2000s, to spend an undergraduate semester "abroad." While other friends spent time in romantic European cities, snorkeled in the South Pacific, or ate *lomo saltado* in Peru, I somehow ended up at a glass dome in the high desert, a structure made most famous for many in my generation by Pauly Shore's unusually bad comedic film *Bio-dome*.[22]

During the period in which Columbia University was wrangled into managing this massive complex, the university developed a campus around it for the study of environmental science and policy.[23] I had read that the complex was being sold to a developer intent on building, in all seriousness, a luxury senior community named Biosphere Estates.[24] I wanted to return once more to the Biosphere 2 campus mainly to reminisce: to see the house that my friends and I lived in (now part of a University of Arizona conference center but otherwise unchanged), to enjoy the view of the Santa Catalina mountains, and to walk around the familiar, scrubby Sonoran landscape and to share this peculiar experiment with my ecologist colleague.[25]

The trip to Biosphere 2 and the ACES conference weren't linked intellectually for me until we went on the "under the glass" tour of the dome during my visit. Our tour guide was a friendly older woman, a denizen of the Saddleback Retirement Resort, part of the twisting tongue of sprawl that extends from Tucson to just down the valley of the Biosphere 2 complex. She joined our group to begin the tour in the "human habitat" of the complex, the former dwelling space for the members of the two "manned missions" of Biosphere 2. Our tour group also included about a dozen enthusiastic husband-and-wife members of a Harley Davidson motorcycle club from Sun City, a retirement resort near Phoenix. Winter is the only time, wardrobe-wise, in which it makes sense to be a part of a motorcycle gang in Arizona, and my colleague and I were the only members of our tour group not wearing leather.

There, in the former human habitat, our guide began extolling the biological and engineering virtues of Biosphere 2. It is an enormous three-acre glass-and-steel greenhouse that aimed to reproduce the basic life-supporting functions of the earth's biosphere ("Biosphere 1"). The structure that encased this ambitious microcosm flexibly appropriated design principles from Buckminster Fuller's geodesic domes. During its two enclosed missions, eight Biosphereans dwelled in and shaped the life within this massive structure that included over 3,800 plant species distributed across five wilderness biomes (rainforest, savannah, desert, ocean, and marsh) and two anthropogenic biomes (for intensive agriculture and human habitation).[26]

Biosphere 2 may have been both the grand peak *and* the demise of a remarkable era of closed ecosystem research that had as a long-term goal the ability to explore and colonize space through the making of a closed system that replicated the life-sustaining function of the earth's biosphere. Despite being a part of this long history of closed system research, the Biosphere 2 project was often dismissed by both the scientific establishment and social theorists. While Jean Baudrillard described the Arizona desert as a "magical" site "which allows the problem of survival to be examined, both that of nature and that of species with equal rigour," he was critical of Biosphere 2's mode of performative experimentation.[27] He called the enclosure a "glass coffin" that "seeks to exorcise catastrophe by making an artificial synthesis of all the elements of catastrophe," comparing life in Biosphere 2 to that of a coma.[28]

A sense of impending catastrophe—destruction of the planet as a human habitat either by nuclear annihilation or resource exhaustion—loomed large in the minds of the Biosphereans. The project was dependent on stable climax notions of ecological systems, in which a productive system stasis could be maintained (figure 2.2). But that stasis in cybernetic models is anything but static—it aims to maintain a balance of desired variables through

FIGURE 2.2 Image of the ocean biome in the Biosphere 2 Center in Oracle, Arizona.

*Source*: Jefferson Eck

ever shifting feedback loops of information. Biosphere 2's feedback loops—the information that coursed within and through this structure—were far more robust and generative than Baudrillard was willing to acknowledge. Baudrillard called it a "relic of the future," but to do so is to discount the degree to which it was the material relic of a social and intellectual past.

After briefly introducing the project, our guide invited us into a consideration of that past by asking us to imagine the kind of planning that must go into creating a life-sustaining closed ecological system. She explained how the Biosphereans had to think thoroughly and systematically through the incredibly complex ecological relationships necessary to sustain just the basic requirements of human survival. The makers of Biosphere 2, she explained to us, "had to take a barren concrete base poured over an air and watertight steel enclosure and build a life-sustaining biosphere from scratch, piece by piece." They had to truck in dirt and select a sustainable web of species—from charismatic mammals to trees to small plants and fungi.[29] The task was made all the more critical because, as our guide mentioned almost in passing (as if this wasn't the juicy tidbit that drew us all there), the Biosphereans were thinking about taking one of these closed systems to outer space—to Mars specifically—to sustain a long-term space colony. It was a vision that they believed had the potential to free humans of what at the time seemed like increasingly immediate Malthusian constraints here on Earth.[30]

She then paused, breaking from her prepared recitation as if permitting us the space to slip toward fanciful daydreams of this enormous desert greenhouse relocated to the Martian surface. After a moment, she broke the silence and directed us to follow her into the biomes. As we transitioned from the human habitat through the airlock and into the oppressively humid rainforest biome, I began to think that the scenario she had painted sounded remarkably familiar.

It was, in fact, almost the precise heuristic that conservation scientists use to explain ecosystem services to audiences who are unfamiliar with the concept. As I reflected, I remembered that I had heard many versions of this scenario before. An illustrative example comes from a Stanford News Service press release that recounts the highly regarded ecologist Gretchen Daily's presentation on ecosystem services at the 1997 meeting of the American Association for the Advancement of Science (AAAS). The subsection of the press release that this quote is drawn from was entitled "An Ecosystem on the Moon" and recounts an early version of what would become one of her classic talks:

> "One way to think about the importance of these [ecosystem] goods and services is to imagine trying to set up a happy life on the moon," Daily said at the AAAS session. "Say you had one year to plan your departure from Earth, what would you bring with you?
>
> Assume for the sake of argument that the moon miraculously already had all of the basic conditions for supporting human life, such as an atmosphere, a climate and the physical structure of soil similar to those on Earth although we know that life plays a powerful role in shaping these conditions," she said.
>
> "After inviting your best friends and packing your prized possessions, the big question would be, 'Which of Earth's millions of species do you need to take with you to make that happy life possible?' "
>
> "Tackling the problem systematically, a hypothetical moon colonist could first choose from among all the species exploited directly for food, drink, spices, fiber and timber products, medicines, various industrial products (such as waxes, lac, rubber and oils) and so on. You might want to throw in some house and garden plants, a few trees from which to hang a hammock or swing, and probably some animals that you wouldn't want your kids to miss," she said. "Even being selective, this list could amount to hundreds or even thousands of species."
>
> "The spaceship would be filling up before the colonists even began adding [many of] the species crucial to supporting those at the top of the list: the bacteria, fungi and small animals that help make soil fertile and break down waste and dead organic matter; the birds, bats and insects that pollinate the flowers; the grasses, herbs and trees that hold the soil in place, regulate the hydrological cycle, supply food for animals and so on."

"Without these, you wouldn't be able to produce crops for food or forests for timber and other products, not to mention the spectacular landscapes that provide intellectual and aesthetic stimulation and lift the human spirit," she said.

Which are these unsung heroes? No one knows precisely which nor even approximately how many species are required to sustain human life. "Even determining which species are critical to the functioning of a particular ecosystem service is no simple, Sunday-afternoon task," Daily said. "Which to bring to the moon? Many of these species have never been subjected to even cursory inspection: no human eye has ever blinked at them through a microscope, no human hand has ever scratched out a name or description of them, and most human minds have never spent a moment's reflection on them," Daily said. "Yet the sobering fact of the matter is, as famed ecologist Edward O. Wilson put it: They don't need us, but we need them."[31]

Quoting the Biospherean John Allen, Dorion Sagan and Lynn Margulis offer a hypothetical from 1991 that is markedly similar: "Imagine for a moment you are building a large ship that will travel through space. The vessel will be tightly closed and must hold everything you'll need to exist for the rest of your life. We can't go alone but must take representative organisms integrated into natural ecosystems with us. If we are to set up an independent colony on Mars, we would have to recreate an Earthlike ecosystem on a smaller scale."[32]

For both Daily and Allen, the most productive means by which to imagine the workings of life as it ideally exists here on Earth was to imagine remaking it on Mars.

This characterization of the natural world in terms of a functionalist hierarchy, in which whole elements of nature could be dismissed for not being useful and in which future worlds are made through an ontology of functionality, caught the attention of Baudrillard. He described the increasingly functionalist ecological worldview as "maleficent ecology" because, through it, "nature—the natural world—is becoming residual, insignificant, an encumbrance, and we do not know how to dispose of it." He continued, "By putting the higher functions into orbit, we are transforming the planet itself into a waste-product, a marginal territory, a peripheral space."[33] In this way, Biosphere 2, the centralized, technocratic enclosed ecology surrounded by desert (though a rather lush desert), is "in its character as an experimental artifact . . . a way of transforming our environment into an archaic residue, to be tipped into the dustbins of natural history."[34]

But Baudrillard's provocative description of Biosphere 2, for which he reserved a level of disdain that approached only his description of Disneyland, perhaps took too seriously the claims and desires of scientists that imagined they could isolate and abstract elements of our material world so effectively that the relations that constituted a function could come to operate in a separate relational sphere. His acerbic account describes what Biosphereans, closed system ecologists, and ecosystem service scientists imagine they can do with the world, not the complex social and historical world in which they exist and have come to produce. Like current critics of ecosystem service science, his critique of Biosphere 2 misses its more hopeful, progressive potential through an uncritical retelling of his target's own narratives of genesis.

## THE ANALYTICAL LIMITS OF NEOLIBERALISM

The striking, but unexpected, discursive overlap is exciting because it suggests a potential opening for considering the political origins of ecosystem service science in a new and more plural way. Political ecologists have been somewhat quick to dismiss emergent market–oriented conservation forms such as ecosystem services as simply another example of neoliberalism. Market-based forms of conservation have been characterized by social scientists as the further restructuring of the material world to facilitate free markets through new forms of "primitive accumulation."[35] Ecosystem service science clearly has connections to neoliberal thought, with its commitment to market-oriented strategies for more effectively distributing a whole range of goods traditionally imagined as circulating outside of the market. Ecosystem service scientists, in typical neoliberal fashion, argue forcefully that the problem of poor resource stewardship is the result of decision-making distortions that result from society's lack of tools for effectively valuing nature.

The ecologist Stephen Polasky, in a white paper on ecosystem services for the Moore Foundation (a major Palo Alto–based funder of market-based conservation projects), paraphrases Kinzig et al.: "'You get what you pay for' (or alternatively, you don't get what you don't pay for)."[36] In a playful, matter-of-fact fashion, scientists such as him concede all projects of distribution to the market. Most conservation scientists don't even describe this approach as reflective of a particular market ideology but merely the empirics of a situation that demands new technologies for identifying and measuring what is most important and valuable to us. It reflects the pervasive presence of liberal market ideology and market metaphors in structuring the way we make sense of a whole host of objects and relationships in the world around us.

As David Graeber has pointed out, one of the most pernicious aspects of neoliberal ideology is the way it creates the sense that there is a complete absence of reasonable alternatives. Neoliberalism is, for all the broad domains its privatizing logic is imagined to penetrate, first about privatizing desire and imagination, a tendency that appears even at times within anthropological analysis, despite an empirical record of work that should guide the discipline in more fruitful directions.[37]

But, of course, there are alternatives. Anthropology and cultural geography speak to both the alternatives to what we might consider a neoliberal system,[38] and the alternative uses and reworkings of what we think of as neoliberal moves or tools.[39] Neoliberalism, it has been pointed out, is rarely carried out in an ideologically pure form,[40] and scholarly attention has turned toward analysis of what Jamie Peck, Neil Brenner, Nik Theodor have termed "actually existing neoliberalisms,"[41] an approach that calls attention to the ideologically variegated and socially rich ways neoliberal projects come (and fail) to take shape in the world. These diverse, "impure" forms of "actually existing neoliberalisms" demand what Stephen Collier terms a "technical approach" to the study of neoliberalism and its diverse manifestations, moving beyond a cursory critique of the perceived form of neoliberalism and turning toward a "substantive analysis of the actual forms taken by neoliberalism."[42] In a related vein, broader critiques of capitalism have argued for the way it is not a singular, inevitable, and evolutionary force; abstract narratives of capitalism fail to account for its heterogeneity and for the multitude of alternative kinds of relations among people, society, and economy that persist within the abstractions of capitalism and neoliberalism.[43]

There is much to be weary of in the various iterations of "market nature" or "neoliberal nature," but there is also a growing empirical body of more nuanced substantive analyses that suggests two things: conservation scientists approach their work from surprisingly diverse political perspectives and imagine their work in ways that run counter to simplistic narratives of neoliberal nature, and the tools that ecosystem service scientists have developed have come to be used and circulated in ways that are surprising and perhaps reflective of the diverse politics inscribed in their making.[44] Shapiro-Garza describes the way payment for ecosystem service programs in Mexico have been "hybridized through multiple sites of articulation and contestation to become a federal subsidy for rural poverty alleviation."[45] Work by Robert Fletcher in Costa Rica and Maron Greenleaf in Brazil suggest similar outcomes.[46] The regularity of these non-neoliberal outcomes of neoliberal projects suggests that the neoliberal analytic has limitations and that a more substantive or technical analysis of these projects opens the possibility for locating more hopeful political threads within their heterogeneity (while remaining steadfastly aware of the potential for these projects and tools to circulate in ways that perpetuate inequality as well). The hybrid or plural qualities of these projects do not simply come into being in the messy cultural space of the field locations of payments for ecosystem service projects but are a part of their original production. Plurality, hybridity, and improvisation are the future of market-oriented conservation projects because they are also their past.

## MILITARISTIC MARKET-BASED UTOPIAN ALTERNATIVES (AND OTHER PRODUCTIVE CONTRADICTIONS)

One way of making sense of this diversity of both production and consumption of ecosystem service tools is to consider the complex, contradictory history of the more specific actors, techniques, and technologies that coalesced during the postwar period to create the larger science of ecological value and planning that we think of as ecosystem services. Ecosystem service science, I argue, is a creative and remarkably influential reworking of a suite of technologies and ways of seeing the world that have their roots in militaristic Cold War governance and the sometimes-awkward transfer of those technologies to innovative private sector uses. The disturbing, and at times perplexing, implications of the Cold War's presence in many new technologies of the supposedly freewheeling entrepreneurial spaces of scientific and technological innovation has been well examined in the STS literature.[47]

Here, I build from their work to place special emphasis on an intriguing, and potentially hopeful, thread within this widely acknowledged link in conservation practice and technology. That is, many of the technologies of a neoliberal nature are in fact inherited from a powerful central state and diverse users that identified with a multitude of utopian, though anxiety-ridden, alternatives to that warring state. In the following section, I highlight more specifically the historical linkage between three highly productive and illustrative social spaces for ecological thought: Cold War–era systems ecology, cabin ecology and closed ecosystem research (Biosphere 2), and ecosystem service science through a narrative focus on the work of people such as Eugene Odum, Stewart Brand, the Biosphereans, Paul Ehrlich, and Gretchen Daily. I will describe the interconnected historical constitution of these ecological fields of inquiry to argue that the scientists working within this space are the inheritors, and potentially the progenitors, of this diverse intellectual assemblage. A more diverse future politics of nature, society, and economy (and the way we imagine that relationship among these discursively distinct spaces) may be dependent on embracing and utilizing this plural history. The elements that are repeatedly understood as destructive internal contradictions in critiques of contemporary ecosystem service science—progressive ecological notions of nature and society, capitalism, and militarized colonialism—are not an emergent contradiction, but a complex and often productive intellectual inheritance.

In a compelling 2006 narrative, Fred Turner offers "a reminder of the forgotten affinities between Cold War research and the ecologically oriented counterculture of the New Communalists."[48] The utopian ecologies of the Biosphereans, through their collaboration with people such as Ed Bass, was "not a story of counter-culture's appropriation by capital; rather it was a story of how "members of the two worlds could come together and legitimate one another's projects."[49] In this collaborative work they came to legitimize each other's ideas and practices in a manner that legitimized more broadly the notion that the planet could be understood as a closed informational system. This notion emerged out of a Cold War–fueled desire for a military command and control system, first through the modeling of a potential nuclear destruction by planners and, second, by computer systems trusted with surveilling and tracking these potential attacks.[50] But this closed-world discourse of the 1950s ironically also permitted an imagination of the world that made ecological models of connectivity, and eventually of ecosystem services, possible.

## THE EMERGENCE OF ECOLOGY AS A DISCIPLINE

What we now think of as a neoliberal reconceptualization of nature was in fact the byproduct of the most intensive governmental support for basic scientific research in the biological sciences in US history (and likely world history).[51] The Cold War era was the height of funding and support for ecological research. Much of the fundamental research underlying contemporary ecology is the product of this period of intense funding and collaboration between biological scientists, the US military, and NASA.[52] In Donald Worster's terms, "dollars and pounds poured into ecosystem research in the 1960s and 1970s like a flooding river onto parched land."[53] The supporting resolution for the International Union of Biological Sciences' International Biological Program urged that "scientists would soon learn how to 'modify the global equilibrium and balance sheet of organic materials and resources, so as to improve the balance sheet by increasing production and reducing losses.'"[54] This phrasing explicitly suggests the linkage between Cold War ecological systems research funding and the emergent notion of nature as a financial resource to be managed. While specific resources—timber or oil, for example—had long been conceived in these terms, this period of research marked a shift toward thinking about entire environmental systems, on scales ranging from watersheds to the earth itself, in these terms.

The historians J. R. McNeill and Corinna Unger argue that the "Cold War world was, in many respects, a scientific world—one in which political, social, and cultural problems were viewed through the lens of science and in which science was believed to offer solutions to the challenges both of everyday life and of international politics, including the Cold War itself."[55] The remarkable power and influence of science during that period, and the redemptive potential it came to offer, was of course the paradoxical outcome of its successful effort to prove its capacity for utter destruction during World War II and during the nuclear arms race of the ensuing period. Ecologists during the

Cold War moved, at times uncomfortably, with the desire to understand and improve life on Earth, but did so with the credibility and funding of an enterprise that very nearly *destroyed* it. The way ecologists concerned with environmental conservation came to produce their own intellectual ecosystems of research and practice that both created and utilized key technologies of the Cold War period demonstrates their ongoing efforts to wrestle with the productive contradictions of their own intellectual histories.

The period after World War II not only created the political and institutional context for advancing ecosystem research but also marked a global period of "monopolization of ecosystem research" by the United States, the product of a preoccupation with reconstruction in Europe and Japan and the fact that the remaining ecologists in Europe "repudiated aspects of ecological theory that had been used by the Nazis and militarists to force conformity on the population and to base racist ideology."[56] The result was that ecology came to take on a distinctly American essence, a science made in and through the social, political, and economic climate of post–World War II. Ecology is in many ways shot through with an anxiety-ridden desire to demonstrate the technical capacity of what were seen as the nation's two great ideological assets: science and capitalism.

In the United States, scientists took advantage of this vacuum and the ecosystem concept's alignment with growing general interest in complex systems theory and cybernetics. It was a strange merger of ecological holism and protracted postwar militarism. Frank Golley highlights the way "ecosystem ecologists benefitted from close involvement in the concerns of American society" by arguing for a role in studying "the base-line ecology of nuclear weapons facilities, the effects of radiation on organisms, and the movement of radioactive materials through food chains."[57] The intensely important and productive relationship between ecology and the AEC meant that most large ecological research teams of the 1950s and 1960s "were centered at the AEC national laboratories."[58] The types of ecosystems that early scientists came to know most intimately were, not coincidentally, those which played host to the largest atomic energy research projects: warm deserts (Nevada test site), cold deserts (Hanford), sandy coastal plains (Savannah River plant), and deciduous forests (Oak Ridge National Lab, Brookhaven National Lab). Shepherded by the growing research program of the influential Eugene Odum, systems ecology came to emerge directly from the earlier field of radiation ecology.[59]

## ECOLOGIES OF INFORMATION: CYBERNETIC NATURE

The productive interface between systems theory in ecology and the diverse field of cybernetics became the boundary space through which AEC-funded radiation research and projects of techno-utopian world-making came to mingle. Cybernetics theory was initially developed by the mathematician Norbert Wiener in his attempt to understand the nature of feedback loops and the way numerous interlinked feedback loops could produce self-correcting systems. His research was initially supported by the military and used to improve artillery targeting in World War II.[60] This interest in thinking about ecological relationships as informational systems was picked up by the Yale ecologist G. Evelyn Hutchinson in his work on metabolism and was built on more influentially by his student Howard Odum and then Howard's older brother Eugene. The Odum brothers would work closely with the AEC to study the way radiation moved through the environment. Radiation then became a form of information that flowed through large-scale ecosystems to elucidate their structure and a material by which to ground cybernetic theories proposed by Hutchinson that many rigorously empirical ecologists viewed as simply too abstract.[61]

The Stanford ecologist Paul Ehrlich came to conceptualize these informational feedback systems in biological and evolutionary terms by coining, with Peter H. Raven, the term "coevolution."[62] For them, coevolution involved "the examination of patterns of interaction between two major groups of organisms with a close and evident ecological relationship, such as plants and herbivores."[63] While their interest in coevolutionary processes was not new (Darwin theorized on these relationships in *Origin of the Species*), cybernetics and systems theory offered a framework for describing and modeling these relationships more explicitly. It also provided a language for articulating classically ecological concepts of relationality in new and broader intellectual spaces. Cybernetics, and its application in systems ecology by the Odums, linked Cold War laboratory research to field biologists such as Ehrlich and techno-utopians such as Fuller—they were all piqued by the potential to model complex systems using mathematics, making some abstracted notion of information a universal currency through which scientists from various fields could collaborate.

Turner argues, "Ehrlich's research and teaching in this period strongly reflected this shift" toward "the study of populations and the principals of natural selection in terms modeled on cybernetic theories of command and control," adding that a "preoccupation with systems-oriented models of the natural world informed both his lectures and the 1963 textbook *The Processes of Evolution*," which moved from hierarchical models of species and subspecies toward "a vision of life as a 'complex energy-matter nexus.'"[64] It elaborated on a longer trend in ecosystem ecology that shifted from the linear trophic models of Charles Elton and Arthur Tansley toward Raymond Lindeman's energy-based models of change, a modeling simplification that made "ecosystems [taken as an ontological reality] amenable to quantitative physicochemical analysis."[65] This shift in conceptualizing and measuring ecosystem relationships made the discipline more flexible to work with institutions like the AEC, thus creating the funding that further shaped and expanded the discipline during the mid-twentieth century.

Despite the military funding origins, the collaborative workspaces that engaged ecologists in the work of atomic energy, like those of Eugene Odum at the AEC's Savannah River project, were actually quite progressive, often informal, and in many ways precursors to the kinds of contemporary open workspaces now associated with the supposedly freewheeling, entrepreneurial culture of Silicon Valley.[66] "New ecologists," like the hippies and communalists of the same tumultuous period, were frustrated by not just the hierarchical models of society but also the entrenched dualisms of Western thought including "mind and manner, actor and action."[67] Their systems thinking attempted to overcome this double inheritance. The tools of cybernetics descended from hierarchical institutions, but they came from uniquely crafted spaces within those institutions that were specifically designed to promote new kinds of interaction and collaboration at the intersection of government, academia, and industry—a collaborative space that Ehrlich's intellectual descendants have retooled through projects such as the Natural Capital Project. It is what Turner refers to as the "forgotten openness of the closed world."[68]

## CABIN ECOLOGY AND THE (RE) PRODUCTION OF SPACESHIP EARTH

While ecologists such as the Odum brothers were engaging in new kinds of government-funded research in collaboration with the AEC, engineers and plant biologists in the field of astronautics were working with military researchers to develop the field of cabin ecology, which emerged as the branch of astronautics concerned with the "environment inside a space vehicle."[69] Scientists attempted to create ecologically closed systems that could potentially be used for shelter either belowground or underwater (in submarines) in the event of a nuclear war. The military was also interested in utilizing space colonization as the potential ultimate "dispersal" technique.[70] Ecologists, including the Odums, became more closely involved in these projects in the 1960s and 1970s through the emergent field of space ecology, an intellectual arena actively produced by them to build institutional and funding relationships between their wing of academia and the growing military and space research complex.[71] Ecology was a pragmatic discipline from the start.

As cabin ecology turned to space ecology and a genuine interest both by government institutions (military planners and NASA) and resigned eco-utopianists, the field of inquiry became increasingly influential in creating the language for talking about environmental problems here on Earth. The Odum brothers' use of the term "carrying capacity," initially a term for defining the maximum number of occupants for a given closed ecosystem, spread in usage.

These projects were also highly influential in developing hypotheses about human carrying capacity on Earth and were seen as a potential vehicle for eliminating Malthusian constraints through the colonization of outer space sites such as

Mars. The work articulated well with the cybernetics-oriented systems theory that was influential in biology and ecology at the time. Ecologists contended that an effective closed ecological system is first and foremost an effectively managed complex information system—the ecology of these closed systems would be in essence a part of the computer: "natural components with mechanical shortcuts."[72]

As Peder Anker has noted, cabin ecology evolved to become a strangely circular (tautological) science for understanding the nature of life on Earth. Cabin ecological designs were initially intended to simulate as closely as possible the complex web of life on Earth in order to sustain human life elsewhere. After being built, however, economists such as Kenneth Boulding and Barbara Ward argued that the modeling would work in reverse such that the "closed economy of the future [on Earth] might similarly be called the 'spaceman' economy, in which the earth has become a single spaceship."[73] Likewise, according to Anker, Ward suggested that the "movement of energy within the space cabin" could be used "to understand life on Spaceship Earth."[74] The abstracted remaking of Earth in outer space became the object through which we were supposed to develop our understanding of life itself. In more mainstream ecological circles, the idea of colonizing space was seen as silly, but the model of life on Earth that was created in the exercise of imagining life in space came to seem entirely reasonable. In this way, the field of cabin ecology came to have wide-reaching influence in ecological conceptions of the relationship between people and their earthly environment. The deputy assistant secretary of the Department of the Interior, in his address to the Symposium on Bioengineering and Cabin Ecology in 1968 said, "There is a close analogy between the ecology of a space cabin and the ecology of the planet Earth...It is no surprise, therefore, that there can be useful information transfer between space cabin engineers and between ecologists and geophysicists."[75]

This notion of a spaceship or space cabin that could be holistically managed gained popularity with the release of NASA's famed *Earthrise* photo from the Apollo 8 mission. That image and those that followed became essential in making the globe a very real object of concern and a mode for imagining it as an object to be governed. These pictures had a great deal of influence on the discourse of the modern environmental movement, as is often remarked, but they also greatly impacted the way models of ecology came to be made and thus how life on Earth came to be imagined and studied.[76]

It was the idea of humans as newly situated external visionaries (in the form of a technocratic manager) that inspired counterculture thinkers such as Stewart Brand to imagine humans as newly appointed gods—living in a period in which vision and technology converged to create the sense that people of this time had a unique capacity to manage and care for what appeared to be a simultaneously more remarkable and more vulnerable world than had ever before been imagined. As Brand stated in the first edition of his soon-to-be hit supply catalog or countercultural literary critique, *Whole Earth Catalog*, "We are as gods and might as well get good at it."[77]

But becoming gods would take effort, research, and a distinctively "scientific approach," for as Fuller famously articulated, the "one outstandingly important fact regarding Spaceship Earth . . . is that no instruction book came with it."[78] He continued: "We are forced because of a lack of an instruction book to use our intellect, which is our supreme faculty, to devise scientific experimental procedures and to interpret effectively the significance of the experimental findings. Thus, because the instruction manual was missing we are learning how we safely can anticipate the consequences of an increasing number of alternative ways of extending our satisfactory survival and growth—both physical and metaphysical."[79]

The Spaceship Earth imagery is still a fundamental teaching heuristic in basic ecology. Odum and Barrett's *Fundamentals of Ecology* is one of the most widely used ecology textbooks. It was originally published in 1951 and has been recently updated in a fifth edition. Spaceship Earth is precisely the language used to describe the developing concern with natural capital. In the text, the authors state, "Unsubsidized solar-powered natural ecosystems can be thought of—and certainly should be highly valued—as the basic life-support module that stabilizes and homeorhetically controls spaceship Earth." They continue, "In these systems, large volumes of air are purified daily, water is recycled, climates are controlled, weather is moderated, and much other useful work is accomplished. These processes and services are termed natural capital."[80] The first edition of the text made mention of a homeorhetically controlled Spaceship Earth, but its link to the idea of natural capital is new though not surprising.

Early ecologists promoting the Spaceship Earth metaphor typically did not advocate for the actual colonization of space. Rather, they thought that the metaphor, which created an immediate sense of constraints and cost-benefit trade-offs, might promote more "sustainable economic and social behavior." Picking up on that idea, contemporary natural capital advocates, drawing on models of ecological thinking that are the intellectual descendants of cabin ecology research, argue that the approach "enables decision-makers to quantify nature's values" and "assess tradeoffs associated with alternative land- and water-use choices."[81]

Despite strong counterculture associations with the idea of Spaceship Earth, the term had a long history in economic and political economic thought. The political economist Henry George first used the term in his book *Progress and Poverty* to describe the critical role of the power and politics of resource distribution.[82] The term gained wider dissemination in 1965 when the US ambassador to the United Nations, Adlai Stevenson, used the term in a speech calling for greater global equity. It was then used in the work of the two economists (not conservative, traditional economists, mind you, but still economists) mentioned above—Boulding and Ward. Boulding used the metaphor of Spaceship Earth to argue, somewhat radically, for an evolutionary rather than equilibrium notion of economics.

Like today's ecosystem service scientists who have drawn upon the space exploration metaphor, these economists, unlike some of the techno-utopianists, felt that the idea of actually traveling to space was an ill-conceived and wildly inefficient use of resources.[83] Like contemporary ecosystem service scientists, the appeal for Boulding and Ward was the potential for this imagery to introduce a model for thinking that promoted more sustainable economic and social behavior. The space shuttle or space colony metaphor creates a materially conceivable model, a highly sensitive space of simulated ecological and ethical exploration. Although utterly inefficient in potential practice, either of them represents a fictional model for thinking that is open to the idealized models of economics.

## SPACESHIP EARTH IN PERIL

Government funding for ecological research was driven by a desire to understand how radiation might disperse through the environment and how, in the event of such violent atmospheric contamination, government officials might disperse people from ground zero. Both aspects of this work were predicated on the imagination of atomic disaster. And these imaginations of disaster came to saturate not just scientific work but everyday life through public service announcements, disaster drills, and even the proliferation of air raid shelter signage, the rusted remains of which still haunt the exteriors of many public buildings.[84] This exercise in imagination, both scientific and popular, and at such a perceived scale, helped bring into being the notion of a "global environment." Even before NASA supplied the stunning *Earthrise* image, the world had begun to hear these narrations of the world's future demise. Hannah Arendt wrote, "Technology, having provided the unity of the world, can just as easily destroy it and the means of global communication were designed side by side with the means of possible global destruction."[85] Faisal Devji, building from her insight, suggests that "the reality of globalization, in other words, is premised upon an apocalypse that has not occurred."[86]

Ecological research then was fueled by the anticipation of a material bomb, but ecologists were increasingly influenced by the idea of Spaceship Earth and became transfixed with a new kind of more dispersed disaster. Capitalizing on the notion of carrying capacity that emerged through cabin ecology and the raw psychological power of bomb imagery, Ehrlich termed this new threat rather provocatively as the "population bomb," which was the title of his globally best-selling apocalyptic treatise of ecological disaster.[87]

Four years later, his impassioned argument for population control was given technical support by the now famous report *The Limits to Growth*,[88] which was commissioned by the Club of Rome, a somewhat nebulous organization led by Italian industrialists whose concern for resource management stemmed from a desire to draw reliably on the raw materials and resources of emergent markets.[89] The organization had a political vision of a world where broad international power existed mainly for securing corporate interests and limiting the interference of emergent third world governments, a sort of global free market governance or global cybernetic neoliberalism.

The report describes a cybernetic world systems model based on just five variables (population, industrialization, pollution, food production, and resource depletion) to project potential scenarios. Called the World3 model, it was the third iteration of a model originally developed by Jay Forester, an MIT scientist and one of the leading early cybernetic thinkers. His student, one of the report's authors, Dennis Meadows, led its development.[90]

While not supplying predictions per se, the model showed a system tendency for a population to overshoot its resources, leading to a collapse of the system.[91] The report tapped into the same apocalyptic fervor that made Ehrlich's book a *New York Times* bestseller but also came under fierce criticism from economists and ecologists, specifically for the kinds of data that were used to drive the model and for the lack of calibration to earlier time periods for which there was data to verify the model's efficacy. With these criticisms, cybernetic thought, and its commitment to decentralization and individualism, ran into an ironic problem: The World3 model was severely limited by the fact that there was a severe paucity of the kind of highly centralized data needed to make the model work.[92] Despite the lack of data, the findings became a global phenomenon and, as with Ehrlich's book, it became a sport of sorts to track the accuracy of the predictions. While people often expressed disappointment in the years that followed that there was no "ecological awakening," only a "turbo response of Neoliberalism," the rapid proliferation of global neoliberal reforms in the years that followed should not be understood as unrelated.[93] The Club of Rome, after all, was founded and led by industrialists who utilized scientific expertise not to save the world but to address the "problématique" of development in a world in which resources were understood to be finite and in which their governance appeared increasingly dependent on fractured, unstable postcolonial states. In this way, the collaboration between scientists and industrialists at the Club of Rome—and the awkward political contradictions and compromises that it entailed—served as a prototype for future collaborative projects of "market nature." These collaborations gain traction through a boundary language of an undifferentiated "social" that needs environmental protection from governments that lack the scope and tools for effectively managing essential resources.

## RETURNING TO BIOSPHERE 2

*Man, who even now finds scarce breathing room on this vast globe, cannot retire from the Old World to some yet undiscovered continent, and wait for the slow action of such causes to replace, by a new creation, the Eden he has wasted.*

—George Perkins Marsh, *Man and Nature*

The delicate, and at times contradictory, interplay between cybernetic ideology and the kinds of technologies and modeling tools needed for understanding exactly what a cybernetic system looked like were manifested most spectacularly in the Biosphere 2 project. It was the grand performative denouement of the era of closed ecosystems research and the provocatively materialized, theatrical response to Ehrlichian narratives of doom.

The sense that the environmental movement was a failure led some ecological and cybernetic thinkers to a different solution than those proposed by the Club of Rome or Ehrlich. In the absence of the kind of effective global governance needed to effectively adjust the variables for survival, they proposed expanding the system, extending the process of human survival and evolution (as both a pressure release valve and genetic reserve) into outer space. While Ehrlich argued that there were only two solutions to the population problem—"birth rate solutions and death rate solutions"—the Biosphereans used the language of Spaceship Earth not as a tool for considering a limit to life within it but as an invitation to explore the possibilities for life outside it. But the work of expanding the space of human evolution first involved creating a microcosm of the processes required for sustaining human life. The grand spatial expansion of evolution required the production of a microsystem that could prove stable in its essential variables. As a result, the Biosphere 2 project seized on the political failures of the postwar conservation movement by offering a provocative alternative that relied heavily on the two technical areas of advancement already described: cabin ecology or closed systems research and cybernetics.

In contrast to the military applications of cabin ecology, Biosphere 2 was interested in expanding this notion toward the production of more holistic livable ecosystems, the type that could support the long-term human colonization of outer space. A surprisingly diverse assemblage of actors in academia, government, and the private sector shared this vision. They combined techno-utopian optimism with a generally pessimistic assessment of human life and governance here on Earth (a combination whose similarity to present-day Silicon Valley culture is not, as Turner eloquently argues, a mere coincidence).

The Biosphere 2 project's entanglement with systems ecology and cybernetic approaches to resource management and future scenario planning then speaks to the unexpected resonances that I encountered when I moved between what I had originally thought of as two separate spaces in my Arizona travels: the large ecosystem services research conference and the Biosphere 2 tour.

I have argued that however different their approaches, and despite Ehrlich's generally nonplussed attitude toward the Biosphere 2 experiments, both Biosphere 2 as the peak of closed systems research and ecosystem service science as the formalization of market nature are dependent on strikingly similar heuristics and very similar ways of conceptualizing human life and the social distribution of resources. Both are also suggestive of the inherent political contradictions that existed within individualistic cybernetic visions—understanding, managing, and successfully acting within this cybernetic sphere was dependent on the privileged, even imperialist, notion of a view from above or outside and thus of people who could be trusted to convey that view and act on it.[94]

## CONSTRUCTING NATURAL CAPITAL

The fact that ecosystem service scientists use the same language as the Biosphere 2 tour guide is not a coincidence—it's the manifestation of a shared way of seeing the world that is the product of a richly complex social history of ecological science. The Natural Capital Project, which is based at Stanford University, is a leading site of research for the development of ecosystem service science and the tools for applying that research in environmental decision-making. Ehrlich advised Daily, the project's director and the creator of the "moon mission" analogy described earlier, for her MA and PhD at Stanford. Under Ehrlich's mentorship, she became interested in addressing big environmental issues by linking a commitment to basic ecological field research with pressing issues such as overpopulation, biodiversity, and resource valuation.[95] This intellectual inheritance is legible in the techniques of contemporary ecosystem service science. Ecosystem services become objects through the merger of a series of interrelated methodologies: basic ecological fieldwork (what Daily calls "taking roll call" of nature), the scaling up of this roll call through remote sensing and GIS-based spatial analysis, and the application of this knowledge about nature's services through scenario-based planning for ecosystem service management using GIS. This knowledge is produced through collaborative work groups that merge invigorating field work with sustained, closed intellectual engagement in built environments specifically designed to cultivate an ecological approach to the production of knowledge.

Ecosystem service science then is a creative appropriation of techniques for seeing the world that aren't particularly new, which carry with them legacies of past ideas and usages, both good and bad. The inheritance is potentially productive, but it can only be realized through an acknowledgment of its history. In my fieldwork, ecosystem service scientists frequently dismissed these threads of relation and instead naturalized ecosystem services as assets that were merely waiting to be discovered or appreciated. Proponents of ecosystem service science argue that "ecosystem services transform natural assets (soil, plants and animals, air and water) into things that we value."[96] It is, in fact, something closer to the opposite: Ecosystem service science transforms things that we value (or perhaps, more appropriately, relations that we value) into natural assets. It takes tremendous social work across various culturally loaded scientific fields to make the relationships of people and things that we value in the world into what appear to be natural assets or, in currently popular parlance, natural capital.

To recognize the socially constructed nature of these ecosystem services is not to dismiss the progressive potential of this work. Rather, it is intended to highlight the spaces within this intellectual history where we might be most successful in locating that progressive potential. By moving beyond the simple and not particularly productive social scientific analytic whereby the work of ecosystem service science is simply a uniform subsidiary of a broad neoliberal project, there is the potential to locate the seeds of alternative ecological and economic futures from within the already growing and well-networked project of ostensibly neoliberal nature.

# 3

# FISH STORIES

## Circulating Narratives of Race and Colonialism

### SECURING VALUE

"We can't *actually* show that people are benefiting from ecosystem services. What *is* an ecosystem service anyway?"

I was sitting across the table from James, the Belizean policy manager for an international conservation organization, in his office suite in a gated, sterile concrete strip mall on the outskirts of Belize City. The shades were drawn over the barred windows; the only light came from an overhead fluorescent strip light. It was a strangely dark, cloistered space from which to imagine new forms of value in nature.

In questioning the ontological status of ecosystem services, James broke character briefly from an at times proselytizing and naturalizing discussion of market-oriented models for valuing nature. His statement articulated the fact that ecosystem services and natural capital are not things simply awaiting recognition but rather come into being only under particular arrangements of people, materials, and technologies—they must be made. And, once made, they must be secured.

In *Wage Labour and Capital*, Karl Marx famously claimed, "A cotton-spinning machine is a machine for spinning cotton. Only under certain conditions does it become capital. Torn away from these conditions, it is as little capital as gold is itself money, or sugar is the price of sugar."[1] Likewise, natural capital and ecosystem services, the concepts by which environmental scientists increasingly articulate the value of our material environment, are not things in the world awaiting valuation as capital. They must be brought into being socially and historically. I argue that, as a result, the value the landscapes come to possesses as natural capital often reflect and reproduce the colonial relationships of social inequality that materially and symbolically produced them. Those landscapes, including the social and political relationships that constitute them, were produced and continue to be maintained through colonial relations of security—relationships that have evolved over time from geopolitical technologies such as the Wyke-Aycinena Treaty through which Guatemala granted the British sovereignty over most of the territory that would become contemporary Belize to the production of reserved "Queen's Land" to more recently the taking of a vast protected area (largely from the remains of the Queen's Land).[2]

"Well, what *is* an ecosystem service on the reef then?" I asked James.

"It's something that the reef, which is our natural heritage—or as we say now 'natural capital'—*gives* that is valuable to the people of Belize. . . . Storm protection, tourism, our fisheries."

"How does it become valuable for people?"

"Well, first you have to identify it. People know something is valuable, and they know to value it if we as scientists can show them it is valuable."

He described the importance of fisheries as an example. "But don't fishermen know that fish are valuable?" I replied.

"Sure, sure, but with these models, everyone will know the fish are valuable and need to be protected. This is our heritage—the heritage of *all* the people of Belize." Identifying and acting to protect the country's "natural capital," its material heritage, was necessary, he said, to "secure the social and economic future" of the country.

In conveying the value of fish through a narrative of heritage, capital, and security, James gestured toward the intersection of colonial logics of security and race with emergent logics of what has been termed "market nature."[3] Belize, like much of the Caribbean, has been constituted socially and economically by a colonial history of racialized dispossession and inequality.[4] It is a landscape whose nature was materially produced through the spatially segregated racialized labor of colonialism.[5] The logic of the universality of money value—articulated through a landscape imagined as natural capital that renders ecosystem services—is based on a curious set of assumptions for a colonial territory in which money value has been historically made by assuming a striking lack of universality in the people that constituted it. The belief in a universally valuable landscape aligns with rhetoric in ecosystem service science and market-oriented conservation more broadly, which suggests that the material environment has a natural monetary value that is simply awaiting recognition for better decision- making. Natural capital as a given object that produces economically valuable goods—here, ecosystem services—then seeks to mimic conventional notions of finance capital, both in its imagined fungibility and in its naturalization of the social relations of inequality that make the accumulation of capital possible.[6]

The valuable fish that James described will be the slippery object through which this chapter traces the emergent relationship between new modes of environmental knowledge and value, and older colonial modes of racialized security reveal themselves in an attempt to secure the value of Belize's natural capital. I homed in on fish in my conversation with James because I had recently become curious about the gaps between how ecosystem service scientists were describing their valuation of finfish (the bony fish that we most typically associate with fish markets) and the material and social relations that made those fish valuable in my ethnographic observations. In creating a model for valuing fish that erased or glossed over much of their social circulation through spaces of racial and class difference—the material movement of fish before and after landing in a market—these models risked naturalizing the inequality that made fish, and so many other objects that were coming to be understood as ecosystem services, socially valuable. In what follows, I trace the material and social relations by which finfish come, in various ways, to market to account for the historically constituted colonial relationships of people and landscape that make the contemporary ecology of Belize valuable. I argue that the narrative gap between the social circulation of fish that I observed and the models of their circulation in ecosystem service rhetoric is representative of the way ecosystem service valuations, often inadvertently, naturalize histories of colonial dispossession and the racialized spatial modes of securitization on which colonialism depends.

The argument is situated at the intersection of work in anthropology that examines 1) the role of environmental conservation in modeling nature as a market object (and the gap between the model and lived social lives of landscapes); 2) the history of a racialized securitization of nature and value under colonialism; and 3) work in maritime anthropology that considers the social life of fisheries under colonialism and the movement between formal and informal in the circulation of marine resources.[7]

I trace the making of finfish—the social and material relations on land and sea by which the fish become an object of concern and value—by moving between two primary ethnographic locations: San Pedro and Belize City. Belize, as it is still sometimes referred to, is the economically struggling former capital. It sits on a narrow, low-lying, peninsula (figure 3.1). Despite its enduring economic struggles and an urban aesthetic that tourist guidebooks dismissively refer to as "ramshackle," the city remains a bustling social center and the headquarters for most environmental nongovernmental organizations (NGOs) and governmental bodies charged with managing the coastal zone. While ethnically diverse in ways that parallel the rest of the country, Belize City has historically been a center of Creole social life. The term carries a range of social meanings in Belize but has been generally been "associated with black identity and mixed ancestry."[8] Creole people often identify as the descendants of slavery in Belize, and through this position they consider themselves the founding people of the independent nation, having an origin in the country that dates to the beginning of colonialism (although this narrative is increasingly challenged by the Maya peoples).

Meanwhile, San Pedro, once an isolated Mestizo fishing village, has grown rapidly into a large and diverse economic center that draws new residents from throughout Belize and Central America. The town is located near the southern tip of the long, narrow offshore island of Ambergris Caye, about thirty miles northeast of Belize City. The island is separated from the Yucatán Peninsula by only a narrow channel and is composed of the limestone detritus of previous reefs. It runs parallel to the current barrier reef, separated by a half-mile-wide lagoon. Inhabited portions of the island are generally a mere three to five feet above sea level. People in San Pedro literally live their lives atop the Belize Barrier Reef and are thus deeply enmeshed in the new ecologies of value being elucidated by ecosystem service scientists.

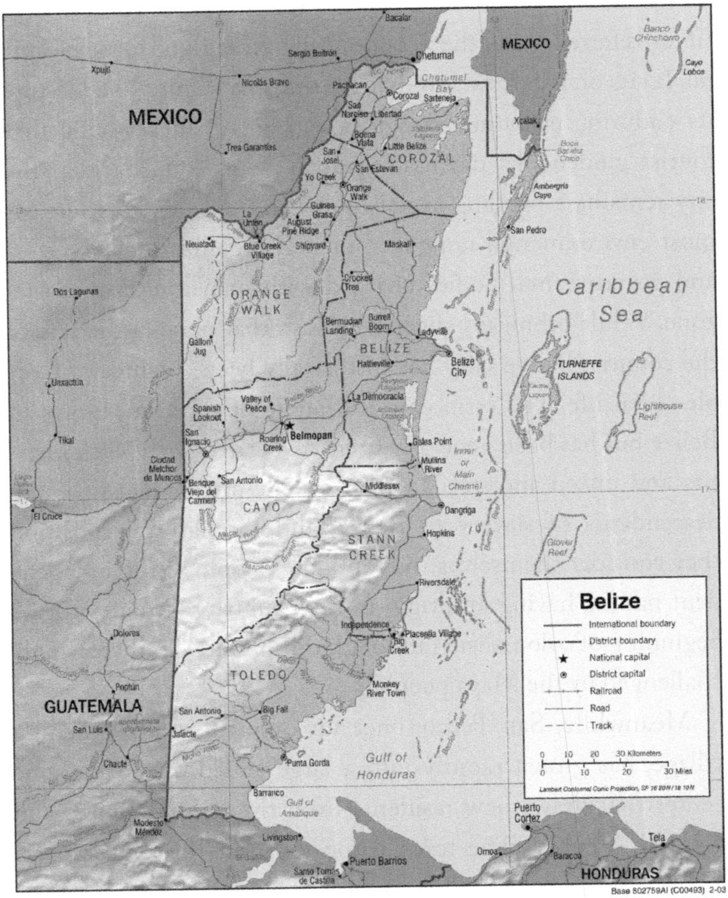


**FIGURE 3.1** Political map of Belize.

*Source*: University of Texas Libraries

Unlike Belize City, San Pedro, since the middle 1800s, has been identified as a Mestizo community. Mestizo is the term now used officially to describe all people who self-identify as being of Spanish descent and is derived from the Spanish term that traditionally referred to mixed Indigenous and European ancestry.[9] The "native" families of San Pedro, referred to as San Pedranos, are generally families that trace their ancestry to Mestizo refugees who fled to northern Belize, rapidly doubling the territory's population, during the Caste War of Yucatán. The protracted, bloody conflict in the neighboring Yucatán Peninsula in Mexico lasted from 1847 to 1901 and quickly changed the ethnic makeup of northern Belize, leading to the first permanent settlement of the colonial period on Ambergris Caye. Today, the town of San Pedro is the rapidly growing center of tourism in Belize, and in the past two decades tourism has become the thriving center of an otherwise sluggish economy. During this period of fast-paced expansion the town has also become more socially diverse with the migration of mainland residents seeking employment, international tourists, and elective migrants, or expatriates, largely from the United States and Canada.

I situate Belize's colonial history in the broader history of the Caribbean to show the way the country's now valuable landscape was made through its distinctive position as a British colonial possession in mainland Central America. The landscape reflects the British colonial effort to secure its possession and resources by producing and managing modes of racial difference. Colonialism and capital in the British Caribbean have long been produced through processes of racialization and the production of racial difference.[10] Furthermore, during the colonial period, Britain repeatedly strengthened its control over capital and territory by offering to securitize emergent nations in the face of racialized threat. Following a similar move by Jamaica after the 1865 Morant Bay Rebellion, the previously autonomous British Honduras Legislative Assembly relinquished a degree of political control in order to become a more secure crown colony in 1870.[11] The move was described by some at the time as a form of political suicide made in the face of what was perceived as an existential threat to the colony posed by the Maya peoples.

During this period, as white British and Belizean Creole people confronted an influx of Yucatán people, they solidified a narrative of "nativeness" that rendered a simplistic binary framing of meaningful ethnoracial difference in the country between the racial complex imagined as native (white or Creole) and that marked as the invasive other (Mestizo, Spanish, or Indian).[12] While Nigel Bolland has rightly pointed out that this dualist framing is a simplistic rendering that doesn't accurately represent the empirical reality of diverse racial and ethnic diversity, Laurie Medina has shown how this simplistic framing nonetheless continues to shape narratives of national belonging (and exclusion) in the country.[13] In Mesoamerica, surrounded by the reach of the Spanish Empire and in a territory that largely resisted Spanish colonization through fierce Maya resistance, this framing of course involved a radical reversal of historical reality. In contrast, however, Ambergris Caye has a history of settlement by both the Maya peoples and then, more permanently, Mestizos fleeing the Yucatán during the war. As a result, in Ambergris Caye, in contrast to Belize City, "nativeness" has been defined traditionally through Mestizo identity, while Creole, and to an extent white, has been imagined as the invasive other.

British colonialism in the Caribbean, including Belize, flexibly produced and appropriated articulations of racial and ethnic difference to render a tense compromise between control, security, and resource exploitation. Belize has long worked through the spatial logic of colonialism by using "'natural' landscapes . . . to limit access to resources, to intensify vulnerabilities and to cordon off subject populations in the name of protection and welfare," practices, Ann Stoler notes, that align with what Michel Foucault has referred to as the "imperative to 'defend society' from its internal and external enemies and to annihilate specific populations in the name of order and peace."[14]

Beginning in the 1950s in the West Indies and spreading to Belize in the following decades, the Caribbean shifted from being a plantation economy to a paradisiacal tourist center and now as reservoir of value in nature.[15] But the shift toward maintaining these imagined geographies of tourism and conservation has not meant the end to a Caribbean postcolonialism period committed to security and order—it has merely shifted that fraught locus to protecting the relations that are imagined to secure a new value in nature and thus the state.

I move between these two locations, Belize City and San Pedro, that hold distinctive positions within the shared colonial system. Ongoing modes of coloniality in contemporary Belize are saturated by a spatialized racial politics, and in this chapter I show how a seemingly simple ecosystem service, the economic value of finfish in a market, comes into being through a set of complex fish and human relations that play out across an uneven landscape. This ethnographic tracing of ecosystem services reveals how these objects become valuable through historically constituted social relations that act to create a landscape that enables, sustains, and naturalizes a colonialist extraction of value under the relatively new framework of ecosystem services.

## SECURITIZING NATURAL CAPITAL

In 2009, the World Resources Institute (WRI) published a landmark study on the value of Belize's coastal capital.

Partnering with conservation professionals there, the WRI estimated the value of "goods and services"—in this case defined rather narrowly as income from fisheries and tourism and the value of "storm protection" from the reef and mangrove's capacity to dissipate wave energy—to be approximately US $500 million per year.[16]

The study reflected broader shifts in the practice of environmental conservation toward market-oriented methods. Partly in response to the critiques of protected areas–based conservation from the social sciences, recent environmental conservation efforts have focused on the protection of ecosystem services and natural capital. Moving toward these forms of environmental conservation is partly in response to critiques of the so-called "fortress model" of conservation dominant in Western conservation practice for much of the twentieth century,[17] but has come to spur a new round of criticism focused on the neoliberalization of nature through increasingly market-driven approaches.[18] Robert Costanza, the ecological economist, defined ecosystem services in his groundbreaking 1997 paper as "the flows of material, energy, and information from natural capital stocks which combine with manufactured and human capital services to produce human welfare."[19] A focus on the protection of ecosystem services, and the natural capital from which they flow, represents a reorientation of the conservation project from producing a flow of protective services from benevolent *people* to some natural space or object and toward a model of a benevolent *nature* providing a flow of services to *people*. This conservation model focuses on protecting the flow of goods and services from nature in a fashion analogous to ensuring the connectivity and stability of the supply chain for a traditional commodity.

The reframing of nature as a form of capital that provides services is in part a rhetorical technique aimed at simply reorienting people's understanding of value, a metaphorical critique of narrowly capitalist models of value. But as these metaphors came to harden and shape the way scientists understood and modeled their object of concern, these commodities became increasingly real to conservationists. Conservation scientists then began collaborating with economists to conceptualize markets in these services. These markets, they theorized, would cause people and institutions to make decisions that took into account the tremendous value of nature.[20] Scientists and policy professionals expressed a similar confidence in the transformational capacity of a natural capital or ecosystem service approach immediately following the WRI valuation study. But in the years that followed, they began to express concern that the valuations were not widely shaping land and resource policy in government.

They did, however, point to the way this new notion of value came to circulate in popular discourse about the barrier reef and its protection. Much of this discourse echoed that of James, emphasizing the need to secure Belize's future by recognizing the incredible value of its natural heritage. But as such claims came to circulate, they raised important questions about citizenship and belonging, the uneven political economy, and the historical production of Belizean nature through the racialized spatial production of the colonial landscape. Part of the difference between the conservation imaginary of "ecosystem service circulation" and the social reality of its making results from a failure to consider the Belizean land- and seascape as a social production and historical artifact that reflects, and often maintains, a highly racialized colonial history.[21] New forms of seemingly raw ecological value are being made through the maintenance of a very old social and political landscape and its uneven regimes of value. As a result, the value of the reef was used to make uneven claims on value and the nation. This was especially true in San Pedro.

For those residents, the valuation of the reef, however, may represent Belizean heritage but also a form of economic value that belonged there and that they felt compelled to protect. The reef and its value as natural capital became the symbolic object from which it could be argued that San Pedro was producing far more value than the nation was returning to it. This attitude was evident at a contentious meeting to discuss a rise in crime that I attended in 2012.

As I arrived in San Pedro for a second period of fieldwork in late spring 2012, these naturalized narratives of racial and ethnic difference were playing out as people throughout the community talked of a crime wave on the island. From the taxi drive north from the airport to my first discussion with the couple hosting me and to chats at a local happy hour, talk of the "crime wave" was constant. That first night back on the island, I biked with my neighbors three miles south to join a group of what turned out to be mainly expatriate white and upper-middle class San Pedrano "neighborhood watch" group members gathered at a weathered resort restaurant to address perceived issues of safety. The police department sent, as their representative, a low-ranking member of the tourism police.

The officer called on to attend and address the crowd said that he was aware of the crime problem. As he talked, though, his tone and demeanor slowly shifted as though energized by the probably unfamiliar forum for voicing his opinion. He said, "As a Belizean, it saddens me to see foreigners being robbed." The officer suggested that he not only knew that there was a crime problem but also that he knew quite specifically its origins. He stated that he was a "Belize City guy," and that he "can recognize the criminal element from Belize City, *here*." These people committing the crimes, he said, are "George Street guys." He said he knows who they are and gives them a difficult time within the law. He was doing—directing this comment to the large cadre of American expatriates in attendance—"what *you* would call 'stop and frisk.'" He also said that he was arresting "suspicious individuals" for loitering if they were seen hanging around hotels (and that they can then be held for forty-eight hours).

Identifying Belize City as the origin for the island's crime problems created an opening for attendees to interject with a very specific but evocative and recurring suggestion for solving the problem: passenger manifests. These are detailed lists, verified by government identification, of passengers for each boat that moved between Belize City and San Pedro, similar to what is produced for commercial air travel. "We don't know who is coming to our island," said one resident, "[and] we don't know who is responsible when there is a crime. They just get right back on the boat and back to Belize [City]." The maintenance of these lists, they argued, was necessary to protect the island, to maintain it as what one San Pedrano hotelier in attendance called "the most valuable place in Belize."

San Pedro's member of parliament was quoted in a newspaper interview just two years earlier:

> Times have changed and it is very regretful because now with the human rights situation . . . there are many people that [are saying] oh you are taking away the privilege of, in particular, when it happens with the Creole descent. That is the first thing that you will hear. But again, I feel that if we have to do that, we have to do it because we have to regain our community. We see many unwanted persons on the island many times and again I have said let us ask the water taxis for a manifest that will probably question the people not really in a rude way but "Where will you be working? Why did you come to San Pedro? Do you have money to sustain yourself in the meantime?"[22]

The answer to the question of where they'll be working or why they came to San Pedro was in fact surprisingly simple in a way that directly contradicted these narratives of belonging: They were here because they were involved in the work of producing the very kinds of value in the reef that the attendees say they threatened. While there were certainly people from Belize City that came to San Pedro to participate in illicit activities (usually petty drug dealing), many simply come as tourists looking for a safe, relaxed, and breezy place to decompress. But, most frequently, people travel from Belize City to San Pedro to work, often in informal spaces, including fishing, the selling of marijuana, or manual labor supporting the tourism industry (property maintenance, beach cleaning, etc.). Rather than threatening these newly articulated modes of ecosystem service value, the informal and undocumented circulation of people from Belize City and other parts of the Belizean mainland was essential to the production of value in both fisheries and tourism.

The disregard for the right of free movement of Belizean citizens and the degree to which that disregard could be shared by socially respected San Pedranos highlights the taken-for-granted way the overlaps among ethnicity, space, and economy form a commonsense understanding of the nation that naturalizes socially produced forms of resource inequality. As the current landscape of Belize is increasingly imagined as a space of flows to be identified and managed for economic and social futures, environmentalists in the country have inadvertently produced narratives that have come to circulate through a local landscape and vocabulary drawn from an uneven colonial past in which value was made through carefully controlling and limiting the movement and circulation of the colony's limited labor pool.

The concern expressed at the meeting for controlling the movement of people believed to be from Belize City was not an anomaly in my fieldwork. People in San Pedro, especially older Mestizo families, frequently described concerns about and critiques of Creole people. Creole administrators of environmental agencies in Belize City who came to the island for meetings and feedback sessions were variously criticized in conversations that I had with San Pedranos for wearing "small, little caps with no brim," "coming in looking like a rap singer," or "making asses of themselves" by wearing the kind of clothes "that Black guys wear." In criticizing high-level bureaucrats engaged in similar kinds of professional work, middle-class San Pedranos revealed how the inclination toward excluding people from Belize City wasn't merely an attempt to exclude the criminal element but more of a desire to maintain traditional social and economic boundaries, an attempt to maintain Ambergris Caye as a proper island. These were boundaries produced during colonialism, but as San Pedro developed first as a prosperous fishing community and more recently as a tourism center, locals attempted to naturalize and reinforce those boundaries to maintain claims to the kinds of value that were now being located there. However, like colonial value, it was made through the erasure of people who were at the foundation of its production.

Traditionally, the "Belize City guy" label used by the officer has indexed a particular conflation of ethnic and racial categorization, that of dark-skinned and Creole. Since even before the establishment of British Honduras as a colony, Belize City has been the center of African and Afro-Creole sociality in Belize.[23] During the colonial period, this linkage between what became its capital city and the predominance of Creoles in the city was sometimes economically advantageous—some elite, urban Creole people served as the bureaucratic functionaries under the British colonial system. Following the movement of the capital inland to Belmopan in the 1960s in the wake of devastating

Hurricane Hattie, however, there was a major outflow of jobs and power, resulting in a sustained period of economic decline.

The economic struggles of Belizean Creoles in the postindependence period has also resulted in out-migration to the United States and sometimes undesirable social remittances in the form of US-based or -modeled gang activity, such that the use of the term "Creole" in Belize (especially by Mestizos and American expatriates) has in recent years come to index a kind of "blackness" that reflects an increasingly American notion of race, racial difference, and racism.

But the officer's othering of the Creoles (and in the process, himself) is in some ways a reversal of the usual process of difference making and exclusion, a reversal distinctive to San Pedro and that moment. Traditionally in Belize, despite fierce political battles over the nature of this identity, being Black or Creole marked you as a citizen, someone who could weave a narrative of belonging that dated to precolonial Belize and the embellished narratives of the Battle of St. George's Caye. On the contrary, "the Spanish," (as Mestizos are often grouped with all non-Indigenous peoples of Central American heritage) have traditionally been stereotyped as the problematic other.[24] In Ambergris Caye and the northern mainland, however, a fierce distinction is often made between Mestizos whose ancestors largely settled during the Yucatán war and the "Spanish"—the largely Central American immigrants who began arriving in large numbers during the periods of political unrest in El Salvador, Guatemala, and Nicaragua.[25] The San Pedranos' origin story—not Indigenous, not through British colonialism, and not the "new" Spanish—created an air of exceptionalism that was propagated by both them and mainland residents who would often remark that San Pedranos "think they're not a part of Belize." In San Pedro, it was the Creoles who "had the fame as thieves and beggars," the invasive others.[26]

## FLUID RELATIONS OF VALUE

Despite being on an island adjacent to one of the world's largest coral reefs, it is very difficult to find fresh fish and lobster for sale outside of tourism-oriented restaurants. This is a paradoxical fact in many of the insular and coastal communities of the Caribbean, which is driven by the high commodity value of fish and the relative poverty of many residents in the region. During my fieldwork, there was no formal, regular fish market in San Pedro. Nor was there a grocery store that sold fresh whole fish or fish fillets. The high-end expatriate-oriented grocery store well known for its uniquely high-quality fresh produce only carried a selection of frozen fish and lobster tails.[27] They could be found hard frozen and frostbitten in a couple of large, poorly marked chest freezers at extremely high prices (US $10–15/lb).[28] The unaffordability of seafood in the Caribbean is part of a longer, broader trend driven by the confluence of robust export markets in certain high-value species, drastically increased tourist demand, and a growing realization that fish harvests in the greater Caribbean have peaked and are experiencing a dramatic decline.[29]

Eventually, if you asked around about buying fresh fish on the island, one name would come up over and over again: Fish. The one man who is a known fish seller goes appropriately by that name, but he is really named Arquel Johnson.[30] In the last few years leading up to his untimely passing, Arquel become more of a public presence on the island. He sold T-shirts bearing his name and image and handed out formal business cards to potential clients. But in my first two years in San Pedro, his presence was somewhat more episodic. He would appear almost daily at your door for a time and then disappear for weeks only to turn up at the sports bar downtown with a cooler full of slimy shrimp and melted ice—suggesting that he may have been out of commission for a while.

When I first met Arquel, I was attempting to trace the movement of fish and sand in San Pedro to understand the social life of two of the objects understood to be critically valuable ecosystem services. When I asked him if I could go out fishing with him, he was somewhat equivocal in his response. I came to understand through subsequent conversations that his hesitance was related to two factors: his methods when he fished and the fact that he had moved into a primarily selling role. First, he used fairly rudimentary, low-cost tools. Unlike larger lobstering operations in which a sailboat or motorboat would bring out several fishermen and their canoes, Arquel paddled, often several miles, by himself in a small canoe and then would tie it up and snorkel alone hunting for lobsters using a Hawaiian sling-style spear gun. His reticence to take me along with him was simply because of the logistical challenge of using just one canoe. Second, Arquel was only an occasional lobsterman despite his intentionally curated identity as a harvester of fish. He spent far more time selling fish from his bike or shared vegetable stand than on the water fishing. He was indeed a fisherman but had found a more profitable economic niche as local tourist market intermediary. It is a role well-examined by the fisheries literature as intermediaries play an important role because they "effectively channel the different market demands such that the tourism market becomes available to the average small-scale artisanal fisherman."[31]

Logistical limitations, weather, and the time demanded to sell the fish prevented him from fishing enough to satisfy the remarkable demand that had developed around his bike-based market. You wouldn't find Arquel at the dock, or pulling a kayak up to the beach, or along the now redeveloped back piers where others occasionally sold fish. You would typically find him near a wooden vegetable stand along the main road just north of the downtown area, or on Sundays he hung out at the pool of a defunct hotel along a sandy stretch of the Boca del Rio neighborhood. If things were particularly slow (or had gone especially well), you would find him, with his bicycle-mounted cooler propped up against a palm tree, sitting in that plastic lawn chair at the no-frills beachside sports bar.

His multitude of roles and movement through the fish supply chain made the origins and circulation of the fish he sold somewhat unclear as well. His fish were typically already filleted, sometimes frozen, and prepackaged into Ziploc bags in portions that were a bit larger than people really wanted to buy but close enough that they'd be willing to reach a deal to buy the whole thing. He usually sold snapper, but since the fish was already filleted there was no way to really tell if it was *Lutjanus campechanus*, the highly desirable species of red snapper that most people imagined themselves buying. And Arquel often wasn't completely certain either. The mislabeling and misidentification of fish is a persistent problem in Belize and especially in San Pedro. While fisheries management professionals often talk about this as a *limiting* factor in creating a strong international market for Belizean marine products, I found that this uncertainty sustains the possibility for a highly profitable local market in the sale of fish for many small-scale harvesters and sellers such as Arquel.

Belize implemented a national ban on the fishing of herbivorous fish (most notably parrotfish) in 2009 because these grazing fish reduce the amount of potentially damaging macroalgal coverage on coral reefs.[32] A 2013 study, however, found that herbivorous fish continue to be sold and that the problem is uniquely bad in San Pedro, where 33–43 percent of all fish sold were parrotfish (and labeled or sold as something else), compared with

2–4 percent in Belize City.[33] This is one of the main contributing reasons for why San Pedro also has a uniquely high proportion of mislabeled fish for sale. According to Courtney Cox's study, nearly three-quarters of all fish sold in San Pedro is mislabeled.

The study authors attribute the high proportions of herbivorous fish and mislabeled fish in San Pedro to especially high demand for premium fish (snapper and grouper) in the country's largest tourism center. This is certainly a factor. But San Pedro, despite the highest overall consumer demand in the country and its location literally on the reef, has what is likely the least formal market for fish. There is no regular, central fish market, no functioning cooperative, and most sales are now direct from individual fishermen to markets and hotels. The incredible demand for fish brought about by the tourism industry paradoxically brought about the decline of social institutions for managing fisheries (such as the once powerful Caribeña Cooperative) because it became so easy and so incredibly profitable to sell directly to hotels and restaurants there. The result now is an enormous consumer market for fish and lobster that is supported by loosely linked fishermen drawing from largely unknown sources.

The economic sustainability of both the fishermen *and* the hotels and restaurants are essentially dependent on the type of radical uncertainty that has emerged around fish—their species and their origins—in San Pedro. If fish being sold as red snapper were correctly labeled, either as illegally harvested herbivorous fish or as lower-value species, then this valuable finfish economy would rapidly shrink. Uncertainty, the bane of scientific inquiry into fisheries produces here new forms of value and new market opportunities for marginal members of the community.

This uncertainty brings the initial paradox back to the fore. If Arquel isn't catching substantial quantities of fish and lobster himself, how is it all finding its way to him? It turns out that he maintained his singular persona by regularly tapping into a network of people—uncles, cousins, friends, and fishing associates—who linked his fish procurement to relatively distant parts of Belize including the Mestizo communities of Sarteneja and Corozal in the north and the Creole center of Belize City to the south. As he told me, "The tourists are always here . . . but the fish isn't. . . . I have lots of fishers in my family, and I have lots of people to sell to."

Both fish and the San Pedro economy are sustained through networks of material and social relations that extend well beyond the final plank of the island's numerous "bridges" (the strikingly connective term used to describe piers) and beyond the protective embrace of the fringing reef. Fish come into being as valuable commodities in San Pedro via social and material relationships that link the island community to mainland Belize City in ways that run counter to the narratives of difference often used to describe the two places.

## GEORGE STREET, BELIZE: THE PERILS OF STAGNATION

The George Street mentioned by the police officer at the meeting is a frequent object of moral concern in Belize. But for all its fame, it is in fact a relatively unremarkable street in appearance—about three car widths from sidewalk to sidewalk, its worn-out and pothole-riddled blacktop extends a dozen or so blocks north to south from just short of Haulover Creek down to Racecourse Street on the largely poor south side of the city. Only a few hundred people *live* on George Street, but the term "George Street people" is frequently invoked to describe members and affiliates of the George Street Gang (reputed to be a spinoff off the California-based Bloods) and the network of family members and relations that are imagined to be dependent on or benefiting from the criminal pursuits of these gang members.

George Street refers then more to a type of person, or way of living, than a specific geographic place, but it is an idea that remains rooted in a particular urban landscape, and a narrative of that landscape stands in problematic opposition to the narrative of flows and function championed by development advocates including ecosystem service scientists.[34] Here, a problematic ecology of stagnation is believed to manifest itself in the culture of the place. The apparent stagnation is a cultural and economic outcome that is made to appear natural.

The south side is frequently portrayed as stagnant—socially, economically, and ecologically. People linger on the streets. Economic activity is sporadic and informal, and the neighborhood (like much of the city, a former mangrove swamp) suffers from back-ups of water and sewage because of flooding. It is a place where the sustaining flows of people, goods, and resources are envisioned as thwarted (though in reality the George Street Gang is reputed to sustain one of the more complex circulations of goods, people, and currency in the country).

As distant as the south side streetscape feels from the sandy streets of San Pedro, these landscapes are long linked by flows of water and time. In a literal sense, one could follow the canal's languorous but enduring flow by boat out of the south side, turning downstream onto Haulover Creek and out toward the sea, where you would pass under the landmark Swing Bridge in the pulsing center of Belize City (which is manually swung open to permit boats to pass) and quickly arrive at the feverish activity of the water taxi dock. The taxis take tourists by power boat to the popular islands of Caye Caulker and, further north, Ambergris Caye. Once out of the south side, then, there is a relative ease of movement from the center of Belize City to the seemingly distant cayes. The sense of distance—aesthetic, social, and economic—is a social production, not a geographic fact.

The relative ease and intensity of this flow of people and goods out of the south side of Belize City was in fact the very thing that increasingly concerned people thirty miles away in the island community of San Pedro. There, the easy movement of people thought to be coming from the stagnation of George Street and the south side of Belize City (whether or not they really physically came from these places) was seen as threatening to other more important flows—the flows of value that were increasingly understood as emanating from the Belize Barrier Reef and, by association, San Pedro. These are flows of value dependent on safe tourists, sustained fish supplies, the beauty of the reef, and the protection of the island's increasingly developed touristic infrastructure.

## BELIZE CITY TO SAN PEDRO: SELECTIVE FLOWS

Later that week, I ran into Arquel at an open-air beachside bar drinking a Belikin beer at the end of a workday that reached past sunset. I told him about the crime meeting and what people were saying about George Street people. I knew he was from the mainland and had family in Belize City and throughout the northern mainland and that he traveled between the island and Belize City regularly. "They think this is an island." Clearly annoyed, he gazed down and then back out across the sea in front of us. "It's not an island."

Arquel described both subtle and explicit practices of exclusion that he felt in daily life—places he was told he didn't have a right to fish, bars where he was not allowed to engage in conversation, and particular resorts at which he was not permitted to sell his fish. He described prevailing threads of nativism that imagined the island as a closed social system, a system within which he and his family were often treated as invasive. By his account, however, he had lived and worked on San Pedro for decades. He was keenly aware that through his labor, a form of resource circulation more complicated than being a mere fisher, he revealed the falsity of such island narratives.

Arquel's concerns highlighted the fact that these emergent narratives of exclusion were in fact the troublesome repurposing of very old narratives to produce spatially explicit forms of ethnic and racial difference. As Melissa Johnson has succinctly argued, "Each ethnic group in Belize, both historically and currently, is associated with particular places within the country. Colonial racial discourse has been a critical part of the mutual constitution of place and identity for the different peoples who live in Belize, as well as a critical part of the ways in which each of the ethnic groups in Belize understand and interrelate with the natural environment."[35]

Efforts to secure the value of the country's natural resources have long capitalized on these colonial notions of space that place the Creole outside of and thus not responsible for, or deserving of, value stemming from the nation's ecological wealth. The assumed spatialized ideas of Creole belonging were made through efforts to substantially restrict their possibilities in the colonial era (Belize's independence in 1981 is recent history) for alternative places and modes of settlement and production.

All of this seemed to incense Arquel, who was typically a mild-mannered person, happier to sit at the beachside bar watching football than engage in political conversation. In his frustration, however, and in the many stories he would tell me about his experience as a non-San Pedrano fisherman there, he offered an entirely different narrative about people, nature and the island, and how these objects came into being—a narrative that questioned both ecological models of ecosystem service flow and the xenophobic narratives of naturalized social exclusion that were imagined to protect those flows.

The mechanics of everyday island life, social and ecological, constantly undermined these narratives of exclusion. Ambergris Caye is an island that is made real, given meaning, and rendered valuable not through its isolation but through its connectivity. They were narratives that reveal the networked ways value is not only created but also redistributed in a subversive manner that responds to a social system that frequently seeks to naturalize a landscape of historically made inequality through racialized colonial practices of securitization.

## MARKETING UNCERTAINTY: (RE) MODELING BELIZEAN FISHERIES

When I told the story of Arquel to David, a reef biologist at a conservation NGO in Belize City, he suggested that this networked procurement by fish sellers in San Pedro might not be unique. He explained how, from his experience researching the accessibility of fish in that city, it appeared that there was significant "leakage" out to San Pedro. Until the 1970s, most San Pedro fish was sold in Belize City, either through markets or later through the agents of the cooperative. Now, with the supply of both fish and regular fishermen in San Pedro proving inadequate, I found that Belize City fishermen are quietly selling products back to San Pedro–based fishermen (or just sellers of fish).

To better understand these dynamics, I began spending time at the city's main fish market, the Vernon Street Market. For the commercial center of a Caribbean country, the fish market is rather unimpressive. There is in fact no market in the structural sense. Instead, at the top of the Collet Canal, just in from Haulover Creek, there is a location where boats tie up at the side of the canal and sell their fish largely from the bow of their boats. This is a locally driven market primarily in finfish. More profitable export marine products such as lobster and conch are generally sold in bulk through cooperatives that have their own processing facilities. At the market, anywhere from a dozen to two dozen boats tie up and sell jack fish, snapper, grouper, and the occasional shark (the fins of which fetch significant sums from the city's substantial Chinese community).

While markets such as this one are often imagined as the "end point" or "landing" in models of fisheries' value, the way the fish and its value continue to circulate in highly unpredictable ways after the moment of landing undermines the presumed nature of ecosystem service flows in Belize. The conservation scientists I worked with in Belize and the United States developed models to track what they refer to as the "landed biomass and net present value of fish and shellfish, distribution of landings and value to communities."[36] But they are dependent on a simple valuation model (intentionally) that imagines a biological sphere and a social sphere and aims to simply represent the spatial extent of the people's intervention into the marine ecosystem and where these interventions occur—that is, where the extracted fish ends up for sale and how much it sells for. These models leave out a good deal of information about the relations between people and fish outside of the short spatial and temporal frame centered on this person/ecosystem interface. The models don't address in detail where fishers come from to collect fish where they do, how the fish comes to circulate *after* its initial landing, and how the fish comes to circulate in secondary markets and spheres of exchange after its initial marketing. It is this largely unmarked, post-official market circulation to unexpected places that comes to sustain naturalizing narratives about space, biology, and labor that are often misleading.

From conversations and observations at the fish market, following David's suggestion, I learned that the fish market didn't always serve as a simple provisioning site for the local distribution of fish (though it certainly does do this). Fishers there were also selling in large quantities to other fishers, or sellers of fish, such as Arquel, who would take that fish by boat back to San Pedro for sale to tourists and restaurants (and occasionally have it sent by air in the cargo hold of the small commercial airliners just below the seats of the wealthy tourists who would soon consume it). There it would be sold and consumed as local fish with a uniquely short commodity chain (the chain's shortness enhancing the perceived value of what the consumers enjoyed). But, in fact, the chain linked this valuable fish back to the languid canals of Belize City and to Creole fishers and fishmongers, many of whom lived in the very south side neighborhoods that were repeatedly portrayed as insidious threats to the value of San Pedro during heated discussions about threats to the island community's social and ecological stability. These men (and apart from the cleaning of fish, the fish market is another space dominated by men) circulate fish products, and themselves through their labor, in ways that run completely counter to their naturalized positions in a country thick with the residue of colonial inequality. In challenging their own naturalized social positions, they also then undermine what are believed to be the natural flows of fish and how it is that they become valuable to people.

This more complex narrative of circulation after the landing of the fish is absent from ecosystem service models of circulation and value. The model that ecosystem service scientists used to describe how a valuable fish is first landed was quite complex. Scientists developed models that considered the effects of changing fishing practices, broadscale economic changes, changes in tourism numbers and practices, and shifts in marine resource governance in relation to sophisticated models of fish life cycles in the reef ecosystem. The model for what happens next, however, is strikingly and problematically simple. To calculate the value of the fish as an ecosystem service, the landed biomass was estimated, along with its market value, and from that the net present value calculated. This number, the net present value, was included with the other elements of the overall ecosystem service value of the Belize Barrier Reef.

For the model and modelers, this was the value of the fish to the people of Belize. The ethnographic material in the following section aims to reveal how such a model is problematic in two ways. First, it fails to capture the complex ecology of the fish *outside* of the water and how the relations of fish, people, and technologies that interact to create its postlanding ecology undermine naturalized notions about how ecosystem service value is produced in Belize and how we might imagine an equitable distribution of that value. Second, it naturalizes a racially and ethnically differentiated social history of the relationship between Belizean people and the land- and seascape and its resources, as well as the way colonial and capitalist value in Belize have always been made through exploiting those notions of difference and enacting them on the landscape.

## MARGINALIZED VALUE

This is an enduring quality of economic and social life in Belize. An important feature of the social and economic history in the country has been creative cultivation on the margins, outside the gaze of state power—sustaining flows of people, goods, and currency that would undermine the narratives of how colonialism works.[37] This work continues in the fish provisioning industry in San Pedro. Socially marginal people collaborate with fishers and sellers in the island's marginalized urban "other," Belize City, to make a place for themselves in a challenging political economic landscape. And in doing so they are quietly sustaining touristic narratives of place and value in San Pedro (and thus for all of Belize) in a manner that couldn't otherwise be sustained as the island becomes rapidly overdeveloped and its local resources intensely overexploited.

But this is not to overstate the agency of marginalized people, Creole and otherwise, in Belize. They operate within substantial structural limitations made through the historical accretion of categories and geographies of difference. One afternoon in August, I met up with David to spend the afternoon talking to fishers at the Vernon Street Market. Near the end of the afternoon, he waved across the market to an older, disheveled-looking man wearing dark, worn-out polyester trousers and a wrinkled maroon work shirt. "That's my uncle," he said. "Let me grab him a beer." We ran into the corner shop and bought his uncle a Belikin stout, as well as two for ourselves.

But we didn't stay to drink them with his uncle. David shook his uncle's hand, clasping his forearm in a display of stifled familial intimacy and wished him well before parting. When we walked away back toward the canal and the frenzy of the marginalized entrepreneurship of the market, David almost apologized to me for buying the beer for his uncle. It was obviously not his uncle's first beer of the day and probably not his last. David shrugged, looked down the canal, and said, "He's not going anywhere. A lot of these guys aren't going anywhere. I figure I can at least get him a beer." It captured the difficult contradiction of this space and more generally the economic margins of Belize. Just off to the side of this unexpected, frenzied movement of people and nature, there is a growing class of people unable to work their way into that current and into the shifting circulation of value in the landscape of contemporary Belize. Their marginalized place has a long history. For conservation scientists and anthropologists to partner with their interlocutors in Belize and elsewhere to imagine and enact social futures that capitalize on notions of ecological value, they will need to first find the history that produced it.

# 4

# GROUND TRUTHING

## The Human Labor of Natural Capital

The bus ride on the Southern Highway from Punta Gorda to the Blue Creek turnoff takes approximately forty-five minutes. Leaving the town, the road edges along a steep, short bluff above the murky waters of the Gulf of Honduras. It is a clear morning, and out the left side of the bus, you can look across the gulf and see the craggy mountains of Guatemala and Honduras. Leaving town, the road continues bending away to the west through scattered villages and across rolling hills that alternate between active cultivation and stark jungle.[1] Then, about ten minutes out of town, the road makes a sharp right turn to join the Southern Highway—one of Belize's four primary travel routes—at a small village referred to, without affection, as Dump.[2] Here, at a junction whose name hints at none of the romanticism often associated with the Maya Mountains of southern Belize, the bus picks up one additional large load of people who have taken smaller buses from various Maya villages to the south and west before turning north into more rural terrain.

North from Dump, the road runs along the narrow skirt of the Maya Mountains. Out the left side of the now fully packed bus, the terrain rises quickly to around a thousand feet in elevation. The mountainous land is composed of a patchwork of green shades from the low pistachio-shaded brush of recently cleared *milpa* (more derogatorily known as "slash and burn") to the deep hunter green shades of older growth "jungle." Out the right side of the bus, the mountains' alluvial skirt hurriedly flattens into a broad pine savanna plain that covers most of the ten or so remaining miles before the muddy rivers of this well-worked landscape find their release in the Gulf of Honduras, a now understandably sediment-choked corner of the usually azure Caribbean.

Jonas and I walked along the thickly muddy but well-trodden path through the pine scrub of a broad plain that leads eventually to the sea but first to the Mennonite community of Blue Creek. The Mennonites are an Anabaptist religious sect that has, over the past two centuries, engaged in a fragmented global migration in search of religious tolerance, moving from Germany to Russia and then forming communities in countries across the Americas. Nearly ten thousand Mennonites have made a home in various parts of Belize over the past seventy-five years.[3] Here in the Belizean bush, Mennonite farmers, dressed in dark trousers and long-sleeve button-down shirts with only wide-brimmed hats to shield them from the searing sun, would pass by occasionally. They rode squarely hewn wooden carts pulled by stout horses with faces that somehow suggested a measure of disappointment with their neotropical assignment. Mennonite dress is strikingly and intentionally out of step in Belize. The broad hooves of their horses and the wooden wheels of the carts that they pulled provided an unexpected mechanism for solidifying the otherwise messy mud track of saturated red clay that penetrated the pine scrub east of the Southern Highway (figure 4.1).

"Nice people," Jonas said after a Mennonite farmer greeted us along the track, "but they breed too much."

**FIGURE 4.1** Mennonite farmer traveling by horse-drawn buggy to the Blue Creek community.

*Source*: Author

Jonas was a white volunteer conservation scientist from Spain in his mid-twenties who had recently completed a master's degree in geographic information systems (GIS) at a university in Europe. He was working for a somewhat indefinite period (having stayed already beyond his initial one-year commitment) with a local conservation and community development NGO in the district that I'll refer to here as the Maya Environmental Trust (MET).[4] MET was one of several that contributed to the larger network of spatially explicit ecological data used by scientists at the Natural Capital Project and by Belizean government planning officials. While the NGO has a long history of collaborative conservation work in southern Belize, Jonas had no relationship to or even familiarity with the country prior to this volunteer position. Like many other European volunteers that I met working with this and similar conservation NGOs, Jonas became fascinated with the Belizean landscape and had plans to stay longer, a change in plans made possible by the country's relatively simple visa renewal process for American and European visitors.

Despite Jonas's relatively recent arrival, in his off-handed remark about the Mennonites, he borrowed a classically Kriol term often levied as a playful critique of people who have too many children.[5] In this sense, he offered a subtly racialized "insider" style of criticism about Mennonites, often positioned as outsiders in Belize, despite his own outsider position. Although he had only been in Belize for a little over a year at the time of our outing, he did not hesitate, however, to make use of the well-trodden path the growing Mennonite community had built. Jonas was wearing the decidedly different but equally traditional uniform of the "outdoorsman" conservationist—waterproofed leather boots, lightweight pants, and a shirt with the dull sheen of quick-dry fabric and more zippers than would ever be used.[6]

We were led on the mud track by Roland, a protected areas ranger for the small but influential conservation organization that Jonas volunteered for. Roland was a Q'eqchi' Mayan man, about forty years old, who had been working with the organization for nearly three years. Jonas had expressed his hope that Roland would both guide us through the bush and provide what he termed "native" insight into the type of land use and land cover that we encountered along the way.

Roland was enormously knowledgeable about the area that we surveyed, but his knowledge was based on a relatively recent encounter with this space through conservation labor. He was from a village about thirty miles away and had moved to this area more recently to work with MET. Despite an impressive command of the recent political history of the landscape that we walked and a reasonable knowledge of the flora that we encountered, it was evident this was not the landscape that he grew up with or one over which he possessed the kind of embodied knowledge that people develop over a lifetime of living in relationship to a place, the kind of knowledge that Jonas at times problematically ascribed to a more general indigeneity. Jonas would frequently ask questions of Roland that implied an interest in his "native" knowledge, as a supplement to what was at that point Jonas's largely virtual knowledge of landscape made primarily through engagement with remotely sensed data (data made from satellite images of the landscape) in his office back in Punta Gorda. During the early parts of our walk, Roland would occasionally push back at the predetermined classificatory schemes that Jonas was using to categorize the land cover at each point. Roland's initial desire to introduce uncertainty—"I'm not sure, could be older trees" or "seems mixed"—eventually gave way in the second hour to nuanced narrations that nevertheless led to a conclusion that fit cleanly within the classificatory scheme: "There are many things happening here, but, yes, 'ten-year scrub.'" Despite Roland's obvious discomfort with the classificatory schemes for describing the landscape being used by Jonas, his presence and knowledge were framed as essential to establishing the legitimacy of the "grounded" truths that Jonas sought in this outing.

The goal of the day was to "ground truth" these GIS-produced maps on land cover (the kinds of plants and vegetation for a specific area). Jonas had produced these maps using only remotely sensed satellite imagery at his office in Punta Gorda. The satellites that provide the remotely sensed data used by Jonas follow unmarked but regular tracks orbiting the earth about five hundred miles above the surface. On this day, we'd verify the interpretation of these satellite images by following a mud track of men and horses, which on its way to Blue Creek traversed a private protected area. The track offered a useful transect of land cover and forest types. As we walked the transect, Jonas would call out, "This is clearly savanna burned last year" or "This is "twenty-meter scrub savanna." Assisting him, I would then take a representative picture, usually with Jonas or Roland in it for scale, and Jonas would record the location of the point electronically for later entry into his computer GIS system in Punta Gorda.

Jonas used a small handheld global positioning system (GPS) receiver to record the points. The receiver is the user segment of a complex military navigation and location service technology that went from closely guarded Cold War–era government utility to standard equipment in suburban minivans and cell phones in a matter of thirty years. The use of GPS technology to locate oneself in the world is now so routinized, so built into our most intimate everyday practices, that the magnitude of the technological feat—and the magnitude of the geopolitical context for even pursuing such a feat—is often lost.

The small instrument held in Jonas's hand detects signals from a network of over two dozen orbiting satellites located approximately twelve thousand miles above the earth's surface. When a receiver picks up the signal of four or more of these satellites, the user's location, within a couple of feet, can be determined in just a few seconds' time. By picking up the signal from additional satellites, it can be determined even more precisely—in some cases to under a foot. The device (the one Jonas was using can be purchased at a camping store in the United States for $110) automatically processes data on the amount of time it takes each signal to travel from various satellites to the user into a navigation equation in order to determine the exact location, elevation, time, and speed of the user virtually instantly.[7] With each point, Jonas records data about the land use and land cover at the location. His georeferenced "grounded" data will be used to test the accuracy of earlier data "remotely sensed" using satellite-produced imagery. Taken together, these grounded and remote data will be used to narrate the history of the landscape and how it should be valued.

## THEORETICAL FRAMING

Jonas is one node in a broad network of scientists, data technicians, and modelers who produce data-rich, spatially explicit representations of value in nature using GIS. The models that they produce are used as part of efforts to price ecosystem services so that, advocates argue, policymakers can make more informed decisions about conservation strategies and land use planning that emphasize the role of ecosystems as "natural infrastructure" or natural capital.[8] The models—digital representations of natural space—they produce are one tool in an emergent science of market-based environmental conservation strategies. Social scientists have noted that contemporary conservation governance is increasingly exercised through the use of these sophisticated spatial technologies that produce spaces of conservation intervention imbued with powerful new forms of meaning and value.[9] These technologies also represent a potent new interaction of traditional state technologies of governance (e.g., maps) with what have been termed "neoliberal" modalities of power (market rationality applied to life broadly).

This chapter makes two interrelated arguments. First, the technocratic digital infrastructure of market-oriented conservation is made through the affect-laden embodied labor of conservation scientists. That is, producing the digital infrastructure for rending nature as natural capital and ecosystem services—tools that imagine nature itself as a form of infrastructure—requires forms of affect-laden embodied labor that engage in close, material ways with the landscape. Second, the geographic labor of ecosystem service science is undergoing substantial changes aimed at making its labor more portable, malleable, and interchangeable. These changes create greater distance between GIS labor and the embodied experience of nature and are alienated from the representations that their labor ends up producing. This situation risks erasing the plurality of ideological and affective orientations to nature and value within heterogenous conservation organizations and institutions and creates the potential for capitalist value to become more hegemonic in environmental discourse and practice.

Empirically, these arguments are made through ethnographic attention to two main spaces. In the first section, I describe my fieldwork with GIS scientists working in Belize, describing with particular attention the fieldwork and follow-up office work with a GIS technician volunteering with a conservation NGO in southern Belize. This ethnography helps to illustrate the way GIS laborers conceptualize their work, articulate their own labor subjectivities, and align conceptions of their work with traditional, romantic Western notions of nature (and their ethical responsibility for it)—ideas that run counter to the imagination of market-based ideologies. The chapter then turns to considering the remaking of GIS labor through a training workshop for workers in the Belizean sugar cane and citrus industries that has been organized by the World Wildlife

Fund (WWF). In this section, I describe the workshop and its participants and provide ethnographic accounts of how participants describe using GIS in their work and how this aligns with WWF's interest in governing the production of major commodity crops in Belize and elsewhere in the tropical world. I emphasize the kinds of skills the training provides and those which are left out to highlight how the use of GIS by technical workers comes to be highly structured, meaning that only certain kinds of data are collected and certain kinds of outputs produced. I then place this training session in conversation with a Natural Capital Project event in California in which the project's chief programmer outlined his technical skills–oriented recruitment effort, showing the way the geographic labor for this project draws increasingly on an undifferentiated pool of computer scientists, rather than from environmental scientists trained in GIS and computer programming.

In this chapter, I argue that the material work of producing the digital spatial infrastructure of contemporary conservation is imbued with both affect and ideology in a manner that fundamentally shapes the production of conservation knowledge, space, and value. I make that argument by attending to the mundane ethnographic details of conservation data collection labor, training, and recruitment within the context of efforts to spatially represent the value of nature in coastal Belize, including efforts by large multiparty conservation collaborations such as that of the Natural Capital Project (described in more detail in chap. 2). I situate this inquiry theoretically at the intersection of studies concerned with the social life of infrastructure, performative models of nature and its value (and the increasingly stratified labor of their production), and the role of affect in shaping emergent forms of capitalist value production.

## PERFORMATIVE MODELS OF NATURE AND VALUE

The global conservation movement has increasingly turned to market-oriented means, or what critics refer to as neoliberal methods, to protect critical resources and habitats. This has involved a reconceptualization of intact nature as an economically valuable form of natural capital that renders services that benefit people.[10] This reorientation has in turn led to the need for new ways of representing nature, its value, and how it is distributed through space. Conservation scientists have frequently utilized GIS to create spatially explicit models of environmental value, including maps that can layer data in a manner that brings into being new relational models of nature. These representations bring into being—literally perform—new kinds of nature and relations between humans and nature.

Drawing on earlier work on economic modeling,[11] critical scholars of environmental modeling argue that these models, like models of the economy or climate systems, are not simply technologies of representation but productive, performative tools of governance that create and manage conservation space and conservation markets in new ways.[12] Models are powerful, symbolic tools that don't simply decipher a landscape but simplify the complex relationality of space in a manner that makes technocratic governance seem not only possible but inevitable.[13] Naomi Oreskes, however, shows how "what we call data," the ostensibly neutral pieces of information that animate models, "are inference-laden signifiers of natural phenomenon to which we have incomplete access."[14] Models of natural capital and ecosystem service value, so central to current conservation practice, are made through these small value-laden signifiers, and as such they are not simply tools for better understanding nature and its economic value but are productive of new kinds of nature that possess and represent previously inconceivable regimes of value. It is not a process of identification but instead of production.

While this is often imagined as a virtual form of production, Donald Mackenzie argues that there is an important material dimension to the production of virtual products aimed at fulfilling the broad-ranging desires of various consumers of the environment—policymakers, scientists, investors, and citizens.[15] "Science and technology," he says, "interact not as disembodied knowledge but as embodied expertise (often via the circulation of people)."[16] That knowledge is embodied both in the material technologies of the conservation enterprise and the conservation scientists who labor to animate and make meaningful the representations produced by these technologies.[17]

Despite this materiality, the embodied labor of spatial modeling, through its iterative processes of confirmation and validation, can become remarkably alienated from the material referent or phenomenon under analysis. Myanna Lahsen has argued that the labor of modeling—the actual work of producing the virtual that Mackenzie argues needs closer study—is primarily to analyze the outputs of models, not the physical phenomenon of nature. Through this work, the virtual becomes the referent, altogether leaving behind the presumed material foundation. Lahsen provides powerful ethnographic accounts of the way climate modelers actually come to conflate simulations with observations and samples with data to the point where Lahsen suggests that the producers of models are often the least capable of perceiving the uncertainties and limitations of their models and tools for guiding interventions in nature.[18] In contrast, my ethnographic work with ecological modelers suggests a keen awareness of the limitations of their models, both as a technical representation of ecological relationships and as meaningful artifacts for capturing the intensity of human relationships with the environment. Their desire to engage in field-based ground truthing of model data, and the narrative work that they engaged in during these outings, is an attempt to reimbue increasingly abstracted, virtual models with notions of affective intensity that give meaning to their ostensibly mundane infrastructural work.[19]

The ability to create these GIS-based models of nature's value are dependent on what Geoffrey Bowker, drawing from Donna Haraway, calls the "modality of implosion"—"to accord every category of living thing a single . . . value, so that policymakers can start the work of determining what should be protected and what should not."[20] This contrasts with what was the formerly more popular mode in conservation, the "modality of particularity"—the compulsive production of lists that seek to document everything that is out there. Bowker cautions, however, that in doing so we may be creating "descriptions of the world in terms of repeatable entities: not because the world is so but because this is the nature of our data models."[21] The absence of features and phenomenon of nature that are "singular in space and time" is striking in the case of ecosystem services models. The model's ability or inability to consume this type of data then heavily influences what kinds of scientific labor and data are valid and valued. It also shapes how scientific laborers come to conceive of their subjectivity as workers in a technical field shaped strongly by ideologies about nature, culture, and value. For the conservation scientists and GIS technicians I worked with, who are at the center of this chapter, the labor of ground truthing is an entry point into a space in between, in which a commitment to the model can be maintained while recalling some notion of "nature" that is imagined as singular and imbued with the emotions of attachment and care.[22] They are participants then in the production of a new infrastructure of and for nature. This digital infrastructure that they produce for representing nature metaphorically *as* infrastructure is more than a technocratic device. It is a digital, material, and symbolic space in which conservation scientists increasingly perform their understanding of the relationship between nature, society, and economy and then enact their vision of the ideal maintenance of and relationship between these categories.

## THE AFFECTIVE DIMENSIONS OF PRODUCING THE INFRASTRUCTURE OF CAPITALIST NATURE

This ethnographic examination of mundane conservation labor is made through an attention to its dual role in making a new spatial infrastructure for representing nature and its value; and, in that process, representing the material relations of the environment as itself a form of infrastructure, as a form of "capitalist nature" made to distribute goods to people. The spatial tools of market-oriented conservation then have a dual engagement with the notion of infrastructure—they develop a digital infrastructure that permits its users and consumers to imagine that material nature is a form of existing infrastructure whose distributive capacities are only waiting to be valued and identified.[23]

This differs from traditional ideas about infrastructure, which often image infrastructure as a built system for altering or distributing elements of nature needed and valued by people or to permit the passage of people, materials, or ideas through material space. Brian Larkin defines infrastructure as "built networks that facilitate the flow of goods, people, or ideas and allow for their exchange over space."[24] Anthropologists take this built infrastructure to consider further the kind of social work that it does.

Built infrastructure then is one means of tracing the relationships between people, and people and the environment, over space. In tracing these relationships anthropologists have come to critically consider the underlying political and social visions that mobilize infrastructural projects.[25] In the process, they have revealed the way infrastructure enacts political visions of the relationship between people and also between people and the material world. Infrastructure then is the physical instantiation of an imagined ideal future, a mapping out of an ideal type of political vision for organizing worldly relations. But as Nikhil Anand, Akhil Gupta, and Hannah Appel have pointed out, there is often an empirically productive and revelatory tension between infrastructure as aspiration and as a site of political failure.[26] As Steven Jackson et al. have argued in the context of cybernetic infrastructure, "From the vantage point of the present, many of the infrastructures that support and govern modern lives, societies, and work practices will appear dull, flat, and still" and only reveal themselves at moments of failure or letdown, in instances in which the visions that lead to their production fall short.[27]

Infrastructure then, rather than mere technical object of conveyance, is an evocative semiotic object that both functions as a technology for distribution and object on which new distributive claims can be made.[28] Rather than a mere thing, infrastructure "can also . . . describe a sensibility: a way of thinking and acting in the world," both for its producers and the users that come to engage with it.[29] The digital infrastructure for spatially valuing ecosystems represents and makes possible new ways not just of representing nature's value but of distributing it as well. The digital infrastructure of nature, therefore, like all representations of nature, conveys a great deal about how its creators imagine social and political life to work.[30] Market-oriented conservation projects naturalize a particular vision of environment, but it is a vision that is in fact made real through "an infrastructural base that organizes both market and society."[31] Looking ethnographically at the production of the digital infrastructure of market-oriented conservation helps to reveal the social ideologies and political rationalities that underlie the incursion of ostensibly calculative technical tools for representing and intervening upon nature.

Infrastructural projects reflect the desire to enact imagined futures based on idealized pasts, and as such the apparently technocratic is often made through affect, desire, and nostalgia.[32] They figure strongly in my engagement with conservation scientists and GIS professionals charged with producing models of nature as productive capital. Social scientists have pointed to the role of the affective dimensions of everyday life and how these "social experiences in solution" come to, in an often "obtuse," "erratic," and underdetermined manner, shape the tone and structure of broader political life. For Kathleen Stewart, "the affective" captures the intimacies and intensities of the way everyday social life is experienced. She cites Raymond William's concept of the "structures of feeling," to describe the way these immanent phenomena "do not have to await definition, classification, or rationalizations before they exert palpable pressures."[33] Ordinary affects are social phenomena that capture the ephemeral relationality of social life and the fleeting sensations by which relations between people, and between people and their environment, come to take on meaning. These affective sensations and relations are embedded in infrastructure in the process of their often-mundane production, creating a technical and material project that overflows with feelings and desires of their producers. Anthropologists and other social scientists have paid increasingly urgent attention to the role of affect in shaping and experience in social life, as well as the potential role of affect in shaping the unspoken foundations of political action.[34]

This chapter joins a broader literature that considers the affective components of capitalist expansion and the role of affective labor in constituting the complex social and ecological relations of the contemporary world in the image of global capitalism. Like the work of cloistered traders or informatics workers, spatial data collection and analysis is a vital form of labor for producing new forms of capitalist value that are often hidden or invisible. Laura Bear et al. have pointed out the problems inherent in the binary framing of affective and immaterial labor.[35] Conscious of this concern, I don't use a notion of "affective labor" here in binary opposition to material labor and the material work of capitalism.[36] I aim to instead point to the way conservation labor has interrelated symbolic and material components that are often animated for participants by fragments of intimacy within increasingly technocratic and market-oriented spaces. Their work undermines a binary between material and affective or cognitive and affective work illustrating the way the required technical and material work of environmental capitalism is sustained by the ability of conservation labor to sustain an affective attachment.[37]

These laborers, I argue, sustain their work through modes of attachment to an idea of nature that their labor may well undermine. Drawing on the open-ended methodological gestures of Stewart, I suggest that attention to these fragments—here through the observation of the mundane labor of conservation scientists—may yield different understandings of how market-oriented conservation produces an infrastructure for valuing nature. The space of the affective is a means by which to capture subtler narratives of landscape and meaning as well. These yet unclassified experiences of landscape capture the sense of nostalgia that "plays in the distance between the present and an imagined . . . past," a narrative that permits the interiorization of a phenomenon that is typically imagined as exterior to or prior to social analysis. Attention to the role of ordinary affect in revealing understanding of relationships past, present, and future reveals the way the affective is a political domain as well. For conservation scientists engaged in spatial data collection and analysis, attention to their affective engagement with the material landscape as they sought to verify the meaning of remotely sensed environmental data reveals the way technocratic models of nature as infrastructure come to be made through a host of micro-ethnographic moments in which the material environment provokes small but meaningful responses in the perceptions of environmental scientists that are entrusted with collecting the raw data of complex environmental databases.

## GROUND TRUTHING: WALKING THROUGH PIXELS

Jonas and I were doing our fieldwork that day as part of a process generally referred to as "ground truthing." In GIS and cartography, ground truthing refers to the process of verifying remotely sensed data (for example, from satellite imagery) by physically going to that site to observe what is actually there. Ground truthing, Jonas told me, was necessary "to make sure that what he was seeing was real." though increasingly the process doesn't involve the material ground. Data accuracy is confirmed by comparing it with other sets of data. The ground comes to be produced then through an algorithmically derived view from space.

Back in the office that afternoon, Jonas showed me how he more typically experiences the landscape we had explored on foot earlier. On his office computer was a series of very high-definition LANDSAT satellite pictures of southern Belize. Because these satellites are equipped to detect light in eight different spectral bands, they are incredibly powerful tools for the remote sensing of land use and land cover. Infrared satellites can detect the levels of heat emitted from different land areas. The colors—ranging from black through the gray scale and into a subtle scale of pinks climaxing finally in a penetrating deep pink—serve as an effective proxy for the successional stage of a forest or field.

These images are referred to as "false color" satellite images in that they do not represent the true perceived color of the landscape under examination but instead are images that represent a broader spectrum of light than can be perceived by the human eye.[38] These colors outside of the visual spectrum are represented in the images using an arbitrary set of color classes. The precise selection of colors used in false color composites can vary depending on what the user is trying to highlight, but for the mapping of land cover change, a combination of near infrared, red, and green is used.

The LANDSAT imaging system takes advantage of the distinctive "spectral signature" of vegetation: "The reflectance [of vegetation] is low in both the blue and red regions of the spectrum, due to absorption by chlorophyll for photosynthesis. It has a peak at the green region which gives rise to the green colour of vegetation. In the near infrared (NIR) region, the reflectance is much higher than that in the visible band due to the cellular structure in the leaves. Hence, vegetation can be identified by the high NIR but generally low visible reflectances."[39]

In false color images, the NIR band is typically assigned to be represented with red; vegetation appears in various shades of pink and bright red. Quickly growing plants, usually new growth, reflect the greatest level of infrared light and thus appear a deep pinkish red, while more mature, slower-growing vegetation appears in progressively lighter shades of pink. New areas of growth—likely the result of the recent harvesting of older-growth trees—appear bright pink, standing out in the imagery as an intensely out-of-place threat to the landscape.

Jonas had been translating these tiny colored pixels, with the aid of a computer program, into a map that indicated the stages of forest that could be found in this area of Belize. He had a remarkably intuitive sense for what he was viewing, a capacity to put these endless pixels into a cohesive narrative of place and space. He would describe the relationship between various pixels through stories—stories about how people used the land, about hurricanes that tore down swaths of trees, about failed resort development projects and bad land deals, and about growing Mennonite families and shifting Maya cultivation practices.

Though the job of a GIS analyst is understood primarily in technical terms, his daily work came to be performed through a combination of his own notion of an inherent story to be told and the creative use of an elaborate technical apparatus to perform these narratives as images. He combined this narrative sense with a strong technical capacity to then develop algorithms that essentially taught the computer program to share his intuition and to thereby automatically translate the pixelated images into land cover maps—to spatially represent this narrative of land. This process of teaching the computer and formalizing his mode of image classification through code is referred to as "training." To train the computer to perform the image classification, the GIS user selects a "training site"—a representative space in which the GIS user can carefully classify the different pixels as a model for the computer. The computer GIS program then associates different types of reflectivity with particular land use classifications and can then replicate this work of classification in other areas. This automated classification is referred to as "supervised classification." The computer learns to autonomously do the work of land classification through careful training by its human supervisor or user.

The system classifies each pixel as a distinct and coherent classificatory unit. The pixel is the smallest graphic unit in the image. In Landsat images, each pixel represents approximately thirty meters squared "on the ground," as a particular land use. Each pixel, based on its reflectivity, is translated into a land cover classification, such as primary forest, savanna, or urban. Through this process, Jonas and the computer GIS program produced a strikingly attractive and evocative color-coded map of the types of land use and land covers encountered in this area of Belize. These maps are important because by comparing them for a particular space through time, changes in land cover (and hence changes in measures such as carbon sequestration potential) can be measured quantitatively through time.[40] Embedded within these quantitative measures, however, are a host of nuanced qualitative judgments that reflect embodied understandings of nature, landscape, value, and change.

Following our day in the field, Jonas would "export" the georeferenced points that we collected as a layer of data for the GIS that he was developing. This layer of georeferenced data (meaning the data points we took are linked to coordinates in landscape space) would, if you can imagine a transparency, be overlaid on the map of land use and land cover that he and the computer had been creating using satellite images of the same space that we walked.

The goal of our fieldwork was to confirm that what we encountered in the field correlates well with what Jonas, and his computer, thought he was seeing on the satellite picture. So, for instance, the first point that we collected on our walk was in an area that Jonas labeled as "one year burn," meaning that it was a grassland area that had been burned within the past year. This area of fresh regrowth should have shown up on the infrared satellite as a bright pink and then shaded on his land cover map accordingly (in this case, he chose yellow). If the first point does in fact fall within one of the areas color coded as new growth on the land cover map, then for this site at least, the accuracy of the satellite image and his system of pixel-by-pixel classification is validated. Repeat this process numerous times with the help of a computer program, and the accuracy of his work can be determined quantitatively.

The tool by which this ground truthing is deemed accurate is often referred to as a confusion matrix (figure 4.2). The confusion matrix "visually represents the difference between the actual and predicted classifications of a model."[41] The confusion matrix is a tool utilized in a variety of modeling applications, but

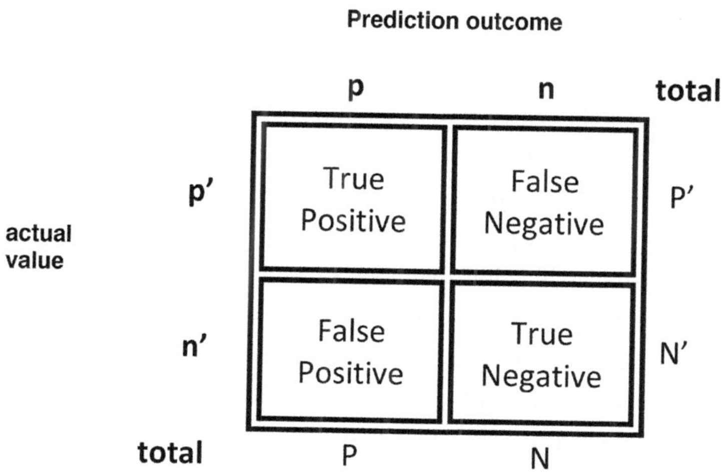


FIGURE 4.2 Simple confusion matrix.

*Source*: Author

in cartographic applications, such as this, the tool is "frequently employed to organize and display information used to assess the thematic accuracy of a land-cover map."[42] It enables a quantifiable comparison of two sources of spatial information and is typically used in land use or land cover mapping to help train the computer program to more accurately and automatically derive land classifications from maps. In this case, the matrix allowed Jonas to compare the guided classification of land cover that he performed to other scientists' classifications of the same or overlapping areas or to compare the computer-guided classifications to what we found in our fieldwork.

The most commonly used measure of overall accuracy for land cover maps is the percentage of correctly allocated cases. But conventional accuracy assessments are based on what is termed the "crisp set theory"—the idea that "the field to be mapped can be divided unambiguously into categories or themes" and "that each pixel in an image can be correctly allocated to a single theme." One European Commission report on methods for land cover classification notes dryly that "these simplifying assumptions help make much of the rigorous statistical analysis described above possible."[43] Noting the limitations of assumptions built into crisp set theory, namely that we can model as though the actual world is composed of a series of discrete pixels, there has been a push toward "fuzzy" or "soft" classification—a mode of classification that allows "for each pixel to have multiple and partial class membership."[44]

Using a standard crisp set analysis of correctly allocated pixels, Jonas confirmed that his classification of the tiny satellite pixels into a land cover type was correct approximately 86 percent of the time (of course, the primary referent is other people's attempts at classifications—it's representation, or turtles, all the way down). The technical literature on ground truthing suggests that this is a strong validation of Jonas's work, but it's the fuzziness of our unpixelated reality that concerned Jonas and other GIS scientists most strongly.

A critique of the binary approach of the crisp set analysis is that all wrong answers are treated as completely and equally wrong. This type of analysis ignores the way each pixel is tied into relations with those around it and the way each pixel's meaning is limited without understanding its context. A wrongly identified pixel might be understood somewhat more charitably with a fuzzy approach that considers that heterogeneity frequently exists even at the relatively fine-grained scale of the pixel and how that heterogeneity within one pixel links it to others (figure 4.3).[45]

The pixels are, after all, arbitrarily brought into being by cameras at random as they whip through space on satellites traveling at 4.735 miles per second. Their spatial divisiveness harbors no biases. But the way GIS analysts and technicians, and the field ecologists with whom they work, make sense of space does. And in the tension between this desire to make meaning of conservation space through their social and ecological engagement with it and their desire to standardize the

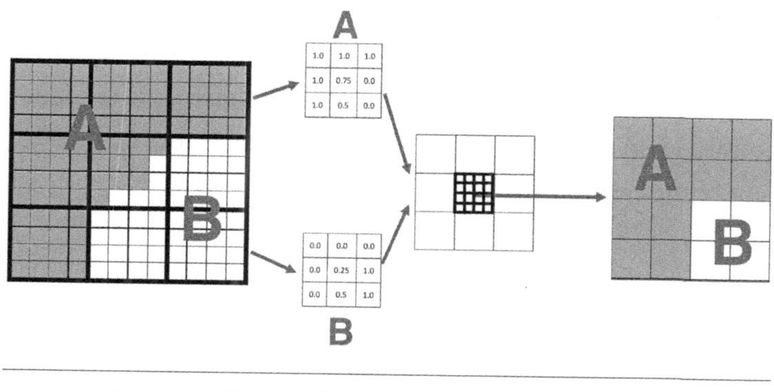


**FIGURE 4.3** Diagram illustrating the mixed pixel problem.

*Source*: Author

representation of that experience through GIS technology lies much of the productive potential of GIS as a political tool for the scientists that use it.

In his ethnography of field ecologists in the Amazon, Hugh Raffles asserts, "An important politics is held in the gap between the utopia of methodology and the awkward realities of field practice."[46] Jonas, and other field scientists in GIS, keenly aware of the frequent need for a fuzzier, more relational understanding of their pixels of analysis, were in some ways better positioned than most others to apply a softer analysis to the broader project of GIS. Jonas enthusiastically consumed critical GIS literature that I would send along to him, literature that highlighted the limitations and assumptions about space and its representations built into the very structure of the software and its rules for making sense of and relating data. GIS, one component of a broad suite of social technologies that come together to produce our images and understanding of nature, does not possess a single unambiguous quality or ethic by which its place in the project of making space, nature, and society can be classified. It is bound up in the infinite relations of contemporary conservation practice. Scientists such as Jonas and his colleagues used the ambiguity that was paradoxically produced through crisp analysis to create new narrative spaces of nature and conservation. GIS's historical tendency toward crisp sets leaves open interpretive space for considering and representing a fuzzier reality.

## ALIEN(ATED) LABOR

Jonas offered his space-age labor in a foreign country essentially for free for reasons that would seem to belie the intended work of the kinds of abstractions that he had become so adept at producing. Within moments of walking the bush path, he told me stories of escaping the city with his father as a child in Europe. He spoke freely of a "nature" that "shaped" him, that "inspired him," of the romantic nature of the modern tradition. It was a notion of nature that stood in apparent conflict to nature as an economic service or form of capital, an image of nature that his data would help to support. There was a sense that his ground truthing wasn't simply an exercise aimed at developing some kind of error algorithm for his land cover mapping. It was an exercise in arresting his own fear that he was no longer seeing, and no longer producing through translation, the thing that had drawn him to the work. He feared that the abstract representation that he was producing in GIS layers didn't adequately convey the object that he imagined himself as protecting and didn't adequately convey how he imagined himself. I came here, he told me in the MET office one afternoon, to "experience something wild, but I spend most days at this desk, in this office." Sometimes, he said, he felt as though "he spent more time with the computer than with nature—it feels sort of strange . . . ridiculous? . . . to come halfway across the world to protect nature by looking at satellite pictures from space."

In his labor and in his thoughtful, self-critical efforts to consider his own subjectivity, Jonas captured what Tim Ingold referred to as the "dialectical interplay" between sphere and globe, "between engagement and detachment, between human beings' involvement in the world and their separation from it."[47] Jonas's experience in Belize is indicative of two related forms of distancing in contemporary environmental conservation. The first stems from the actual material technologies of representing nature and the kind of field of experience that they render for conservation practitioners. Conservation organizations, and many of their workers, operate nearly entirely in the realm of the image—a suite of images that create a totalizing illusion of acting from above (what Jean Baudrillard has called the "hyperreal").[48] Second, this implies new labor subjectivities in which members of the GIS community more explicitly sell a particular kind of labor, in which the product of that labor is imagined to be, in the Marxian sense, forever disconnected from the laborer, the relationship of production erased.

But the fuzziness of their daily practice, and the role of affect and sociality in shaping the way GIS technology is mobilized to produce narratives of the landscape, suggest the way their labor comes to constitute the material set of relations, technologies, and images that produce contemporary nature. In her analysis of Wall Street investment bankers, Karen Ho argues that the everyday practices of investment bankers "make" the market that they are ostensibly acting within.[49] She argues that they make that market through modes of action shaped by habitus, structured ways of experience, seeing, and articulating the world that reflect the socialization of Wall Street more than any inherent quality of the market within which they are supposed to be acting.

Similarly, conservation scientists and GIS analysts are not describing a value in nature that already exists and simply needs to be identified with new tools. The tools that they use—including GIS—bring nature into being as a form of natural capital that can be valued for producing ecosystem services. Market nature comes into being through the everyday work of conservation scientists such as Jonas who shape the image of valuable nature through their own socially conditioned way of understanding the relationship between society, nature, and economy. The actors that engage in this work, however, are not homogenous cogs in a representational machine. Jonas, and other GIS analysts that I worked within Belize and the United States, brought to their work a habitus shaped in different experiences with and orientations toward nature than the more crassly economistic ideal type of descriptions that circulate about the production of market nature. The daily practices of scientists such as Jonas create an image of valuable nature that often deviates from the neoliberal ideal of a fungible value in the material world. Their labor creates the images of market nature, but they are images that in the end embody the more heterogenous notions of nature and value that conservation laborers bring to their work.

Their capacity to imbue these images with heterogenous notions of nature and value, however, is limited. GIS technicians and peripheral analysts are truly a conservation labor commodity, exchangeable and fungible, useful in new places without the challenge of needing to actually know the place—permanently foreign and short-term and often in structurally difficult positions from which to enact a philosophical shift within the project of conservation or even their organization.[50] Their work then leads frequently to a sense of alienation that produces what was repeatedly described to me as a form of anxiety about the circulation of the images that they produced and the manner in which their own labor came to circulate in a fashion that minimalized their social connection to both the image and the place that it purports to represent. As one GIS technician explained to me, "Sometimes, I think I followed my passion into environmental conservation, but sort of feel like a digital widget producer," suggesting that the images he produced were akin to fungible material commodity items, comparing the product of his specialized labor to the classically generic textbook language for an alienated commodity.

## RECONFIGURING THE LABOR OF MAPPING NATURE

I had first met Jonas and several other conservation professionals using GIS in Belize two months earlier at a workshop that he facilitated on behalf of World Wildlife Fund. The workshop was held in Orange Walk, one of the two principal towns of the northern mainland. With approximately thirteen thousand residents, it is one of the largest towns in the country and a center of sugar (and rum) production. The country is only about 180 miles in length from north to south—about the distance from Baltimore to New York City. But the changes in that stretch—topographically, climatically, biologically, and culturally—are quite vast. Northern Belize is starkly flat. It receives about half the rainfall of the lushly tropical south and the people here are predominantly Mestizo.[51]

The workshop was held in the upstairs conference room of the Belize Sugar Cane Farmers Association Orange Walk building. The workshop was primarily targeted at the managers of sugar cane farms so that they could learn to more efficiently and strategically apply fertilizers and pesticides by incorporating spatially explicit data about variables such as rainfall, runoff, slope steepness, and soil moisture into their application planning. The WWF supported this work as part of their efforts to reduce pesticide and fertilizer runoff in the Mesoamerican Reef watershed system. WWF scientists felt that by targeting the four largest agricultural sectors in terms of land area and potential coastal water inputs—sugar cane, citrus, bananas, and shrimp aquaculture—they could substantially reduce the impacts of chemicals and excess nutrients on the reef. It was brought through a broader global WWF strategy to work with business to transform markets so they might align better with the organization's conservation priorities.

The workshop was introducing farm managers to technologies and a means of visualizing the landscape that has historically been used in Belize for more narrowly conservation-oriented purposes. It exported a conservation view of the Belizean landscape to these industrial farm managers. Jonas demonstrated the basic workings and capacities of the tool with exercises that showed the entire country of Belize and how its various watersheds created a visual link between the landscape of sugar production and the socially distant world of the barrier reef. From this broad introduction, the workshop shifted in the afternoon to more specific tools for modeling the efficient application of agricultural inputs as the farm managers worked to individually locate their farms within the broader map of Belize.

Most of the farm managers in the sugar cane industry identify as Mestizo, and as with many of the Mestizo residents of northern Belize, their primary language is Spanish; during casual conversation before the workshop and during breaks, most attendees would speak in Spanish. The noted exceptions were the two men representing the citrus industry, based in the Stann Creek Valley between Belmopan and Dangriga in the southern part of the country (but north of Punta Gorda and the Maya Mountains). The citrus industry has been traditionally worked by Garifuna men from the Dangriga area. While many Garifuna people in southern Belize actively speak the language (which is part of the Arawakan family of languages), the language of business in the citrus industry is primarily English.

In Belize, English is the official language of government and business, and it dominates in most formal settings despite the fact that, by some estimates, Spanish is now the native language of over half of the country's residents.[52] However, because of the relative paucity of exposure to formal Spanish through education and media, the Spanish spoken in northern Belize has a distinctive vocabulary and "many innovative phonological and grammatical patterns."[53] The combination of cultural norms against conducting official business in Spanish, especially when the crowd includes people from other parts of the country, and a lingering discomfort among many northern Belizeans regarding the informality of their variety of Spanish, meant that the workshop led by a native Castilian Spanish speaker from Spain was conducted in English.

The result, in this case, was a challenging program of translation in which Jonas, in a thick Castilian accent, delivered the highly technical English instructions to Mestizo Spanish-speaking men, while the two Garifuna men would frequently exchange glances or roll their eyes at the often-cumbersome translations. The main PowerPoint-driven English portions of the training couldn't fully convey the details of doing the GIS work on each individual's laptop (and each machine—some personal or work computers, some borrowed for the event—was itself slightly different, adding to the challenge). As a result, between training segments, or modules, Jonas would move around the room comfortably and effusively speaking in Spanish to the Mestizo participants, quickly ironing out areas that needed further clarification.

But these interactions also highlighted a challenging reality of the translation of GIS—much of its language is not translatable from English technospeak. While GIS is often translated to *sistema de información geográfica* (SIG) in formal literature, many Spanish speakers, especially international researchers, call it GIS. Other English terms such as "software" or "raster" are not typically translated. Many verbs used to describe technical actions in GIS were simply translated through the creation of neologisms that merely Hispanicize English terms for which there is no obvious translation.[54]

At times, then, it was unclear whether the difficulties in conveying the content of the training were based in the fact that the training materials were presented in English or that they presented a complex new technological language. It was of course both, to an extent, as well as the confusing interaction of these different forms of translation. Most of the workshop attendees had no previous GIS experience, so Jonas needed to balance his desire to provide a theoretical foundation to the work with the funder's goal that these managers develop enough of a foundation that they could carry out the specific technical work related to its potential uses in pesticide and fertilizer application. The sugar cane farm workers were not going be trained as GIS analysts or to even develop a comprehensive understanding of it as a tool and system of knowledge. Rather, they were being trained very narrowly as technicians to use tools in predetermined ways toward predetermined ends.

The predetermined goal, reducing the runoff of excess pesticides and fertilizers into the Belize Barrier Reef is obviously not a bad objective. But the way GIS, historically a tool for conservation in Belize and being introduced here by WWF, came to be circulated into this new space is important for the way it exemplifies a broader shift in the pattern of GIS usage and training. It is a shift that is in one sense democratizing; it is opening up the capacity of GIS tools to a broader class of users, but in the move toward deskilling and rendering technical much of the on-the-ground GIS labor—even, in this case, moving the GIS work of conservation outside of the actual professional field of conservation—it also obscures the social production of this technology and severely limits the flexibility and ingenuity with which the tool can be utilized by this expanded class of users.[55] This lack of space for creative repurposing, I argue, is rapidly undermining some of the surprisingly progressive and transformative potential of its technology within mainstream conservation practice.

## THE GIS LABOR HIERARCHY

This developing hierarchy within GIS work, and the intensifying stratification within this hierarchy, is increasingly well-known within the industry, and this emergent structure of labor was often talked about in conversations with workers and conservation organization leaders. However, because of the computer-driven nature of the work and thus the physical similarity of the workspaces and the machines with which people work in different positions within this hierarchy, it is not always easily observed from the outside. My initial failure to perceive the social significance of this heightened differentiation of GIS labor momentarily disrupted the relaxed tone of a meeting over cold Belikin beers on the shaded patio of a bar popular with the Belize City policy and NGO scene. In an uncomfortable technosocial faux pas, I was speaking with two visiting white American team members, Nathan and Alyssa, of the Natural Capital Project, a collaborative research and applied conservation project with partners at multiple US research universities and two of the largest global conservation NGOs, including the WWF. The project has various ecosystem service modeling and valuation projects throughout the world but had an especially keen interest in modeling and valuing the marine ecosystem services of coastal Belize.[56]

Responding to a description by Alyssa of how the scientists imagined their project not just as technical map-making but as social science as well, I said, "So, you mean not just the work that Nathan's doing as a technician, but—"

My question was gently interrupted by uncomfortable laughter from Alyssa. "He's actually an *analyst*."

Nathan looked toward me, tilting his head, somewhat uncomfortable with the correction, but also clearly appreciating that Alyssa had created an opening to clarify the nature of his work. Speaking almost to each other at this point, they went on to describe the kind of work that they do, to explain the way it differs from how others describe what they do, or even what they trained or signed up for—that their work on protecting nature's function is social and personal. Their work was based on the development of a capacity to, as Nathan put it, "read the whole landscape" and to create tools that helped them train other people to develop a similar capacity. He was not a disinterested producer of technical images; he felt that he was a skilled interpreter of a social landscape.

In articulating these spaces of difference, they were also articulating the perceived agency of their position within the hierarchy of conservation labor. The flexibility and creativity to define their sphere of work and influence defines the social position of their work and emphasizes how it differs from the labor of a mere technician, in which creativity is thought to be limited by the programming work of analysts in the way that a Rubik's Cube permits but also explicitly structures the creative solutions of its users. The official distinction between these two forms of labor is relatively recent. The Department of Labor Employment and Training Administration (DOLETA) "established two new GIS occupations—geographic information scientists and technologists and GIS technicians—in late 2009,"[57] part of a larger consideration of the growth and evolution of GIS labor in the early 2000s.[58] And the way these distinctions are manifested in complex multiparty conservation projects in Belize cannot be disentangled from the racialized relations of enduring coloniality. Nathan and Alyssa, key team members for the Belize project, had, at that point, spent only a handful of days in the country, yet they occupied a privileged position in mapping meaning and value on its landscape in no small part because they were white American scientists.

GIS work is often touted as an emergent and growing professional space that allows people to be trained in a widely applicable and easily transferable tool that is desired across industries. This is largely true. However, there has been considerable debate within the geography community as to whether GIS constitutes a new discipline of inquiry or a new profession entirely. Most geographers argued that it was not a new discipline but that the growing field constitutes a new profession.[59] The division between the disciplines using GIS and the professional field increasingly tasked with developing the tools and even collecting the data for those disciplines is building an ideological firewall that separates affective concerns with the project from its technical renderings. People entering the industry are not just encountering a growing field but also one that is being substantially reconfigured. As GIS became a more recognized field, degree and certification programs proliferated, training a growing number of people more narrowly in the technical skills of GIS. Its earlier use was primarily by nonspecialists who worked in some other field of inquiry and came to use GIS as one tool for approaching that broader inquiry. One result has been that more basic technician jobs are increasingly either outsourced *or* replaced by free web-based applications or more user-friendly tools. These tools are produced by "higher end" labor that usually involves combining strong GIS skills with advanced programming skills. People seeking careers in GIS are advised to "specialize or hybridize"—to become an expert in a specific professional field that utilizes it or become an expert in high-level programming skills that are broadly applicable across fields.[60] Failure to move toward one of these sophisticated poles could result in a GIS laborer becoming a mere "cog in the conservation machine," as one geographer described it to me.

The traditional hierarchy of GIS labor (in a simplified and idealized form) is frequently imagined as a typical pyramid of predictable upward mobility. Workers begin as a technician responsible for doing basic spatial database management and map production. This is followed by a position as an analyst and then a senior analyst. In the analyst position, workers engage in the more complex, and potentially exciting, labor of map production and sophisticated data analysis—more directly producing the kinds of compelling images that emerge from spatial data analysis. These workers typically have a bachelor's degree or higher training specifically in geography or a related field. The top of the hierarchy is the GIS Coordinator or Manager (and even occasionally now, in corporate parlance, the Geographic Information Officer).[61]

There is increasing stratification within this hierarchy. In order to attain the higher-level analyst or senior analyst positions, workers must, beyond basic technical skills, possess extensive programming or "geospatial customization" skills. They are expected to be able to custom-program GIS tools for the needs of their specific organization or industry. This can be seen at the Natural Capital Project, where GIS analysts aren't merely performing creative analyses and map productions but are creating customized platforms that the project will use for sharing their information with the public via their website.

These customized mapping platforms are part of an effort to "liberate" the project's data and planning tools from the expensive ArcGIS software suite whose exorbitant licensing fees are prohibitively high for many potential users of the technology.

The high licensing fees mean that collaborative projects that use ArcGIS as a platform for sharing their data risk seriously limiting their potential "public."[62] While efforts to liberate their tools reflect a well-intentioned effort to broaden the public of their project, they also have the potential to change the political valence of the new tools being produced. Unlike GIS laborers trained in the biological sciences, programming-oriented GIS scientists have less of a stake in the politics of nature and conservation, and there is concern that the tools they produce may come to reflect this. More specialized GIS tools for a project of this kind have the potential to limit the spaces for creative interpretation and modification by ideologically diverse project laborers previously afforded by the somewhat clunky and more generalized ArcGIS software.

## CONCLUSION: "TO MAKE SURE WHAT I'M SEEING IS REAL"

The increasing programming sophistication of GIS analysts and the interest in developing widely usable mapping platforms has been a great boon for the general public and for small conservation organizations interested in understanding spatial relationships but lacking technical expertise. However, this combined stratification and specialization means the self-described "indispensable cogs" who perform day-to-day GIS work are now at a greater emotional and ideological distance from the analysts and programmers who structure the data needs they fill. The result is that the productively fuzzy territory at the boundaries of categories and classifications may become less imbued with the spirit of workers who have more socially and ideologically diverse notions of the environment and the place of their labor within it.

These technically adept programmers are in simplistic terms the ones that followed the "hybridize" path mentioned earlier. But to hybridize as a laborer is also in many to "standardize" the practices by which you labor. At a presentation to Bay Area political and business leaders by the Natural Capital Project, one of the lead programmers for the project made an impassioned call for this type of worker, saying that the project welcomed "smart people" who "love programming" (not nature) and want to "do good."

"Why not," he asked the Silicon Valley audience, "do tech work that matters?" He was asserting that their standardized skill set could travel meaningfully through different fields of practice. While I wouldn't necessarily argue against talented Candy Crush programmers seeking more socially meaningful places of work, the depoliticization of labor critical to some of the most important spaces of political contestation (namely a potentially progressive environmental movement) may in the end weaken the capacity of GIS labor more generally to do good.

Of course, it feels somewhat counterintuitive to account with some remorse for the loss of a more romantic engagement with nature that anthropologists elsewhere rightfully critique as narrowly Western and frequently exclusionary. But I am emphasizing, as Raymond Williams and William Cronon have argued eloquently, that our ideas of nature reflect our culture, and that in losing more diversely derived ideas of nature, we also lose the diversity of cultural ideas that permit us to construct the world around us in lively, surprising, and productive ways.

Jonas's attempt to constantly push his practice of GIS back into the messy field, where data always overflows and where pixels seem mostly irrelevant, came across as an effort to make the seams that tie GIS to the broader world it aims to represent more visible, more vulnerable, and more open to remaking and reimagining. And in that space of vulnerability, of messiness, I'm arguing is one of the key spaces of political alternative and contestation in a modern environmental politics that is increasingly portrayed in deterministic and neoliberal terms.

In his account of ecological research in the Amazon, Raffles discusses the challenges that the field of ecology faces in conceptualizing the forest as a laboratory, highlighting that "it is only through some very specific exclusions that [the forest] is able to function as a site for data generation at all."[63] There is a history in ecology of using "laboratory" forests—or experimental forests—as imagined abstractions, pieces from which they could speak of the whole—Hubbard, Coweeta, Fernow, Harvard. These experimental forests speak to ecology's greatest potential as an experimental science and the greatest anxieties in the socionatural work of making that science real. In a sense, GIS work has swallowed, condensed, and made simultaneously invisible but still palpable (that is, largely unseen but often deeply felt) the anxieties central to the relationship between ecologists and their classical subject, the forest. Turning the forest into a digital abstraction made through the deeply, socially mediated, technical project of perceiving imperceptible light, mapping those reflections in space and making distinctions between the meaning of those reflections is the work of making the vision of the experimental forest scientifically real by distancing it from the "instability and excess" of the field.[64] It is the image of a forest that appears to lend itself greater methodological and even political purity.

But for scientists such as Jonas, there is potential in the instability that remains bound up in the flat, remotely sensed images of the forest. In the spaces of uncertainty within GIS representations—the spaces of judgment calls, where forms of analytic habitus precede the formal articulation of classificatory schemes on land cover maps—Jonas and other geographic laborers possess the ability to saturate their images with interpretations shaped by their embodied social experience of the places represented. In this sense, each pixel is a symbol for a small piece of the forest, and Jonas, as he validates the remotely sensed land use data, has the final say on the relationship between sign and signified conveyed in these maps. This degree of freedom in assigning meaning to the pixel, each a symbol for a tiny fragment of land, has the potential to make the seemingly rote work of GIS labor an important site for sustaining and even cultivating other ways of seeing landscapes in a political space that is increasingly enamored with beautiful but highly routinized images of ecological space.

The way that Jonas performed this fieldwork—stopping in at a small Mayan village, traversing a road made by Mennonite farmers, enlisting Roland to "guide" him, and inviting me to join the event so that he could perform his narration of those relations—was a means, intentional or subconscious, of engaging his technical work with the network of relations and ideas that gave his work meaning. On this trip, Jonas used the technical arguments for ground truthing as a means for producing an entry point into a broader practice of meaning making for the landscape, for his organization, for me, and perhaps most of all for himself. As Jonas talked about his work with me while walking the muddy track to Blue Creek, he revealed a perhaps more powerful impulse that drew him toward ground truthing. That was the desire to simply go outside, to experience as "real" the space that had through his labor been reduced to consuming and producing as a distanced digital abstraction. We went out, he said, "to make sure what I am seeing is real." In saying "what I am seeing," I think he was first referring to what he saw on the computer, but after spending the day with him I also came to hear that as a broader mandate to see the world as he feels it really exists, in space and value, and to overcome what he felt was a disciplinary fixation with abstractions. These abstractions, to universalize notions of nature and its meaning, have made that meaning feel more tenuous for people like him.

Ground truthing wasn't simply about assuring a flat one-to-one correspondence in the coding of a particular geographic coordinate by using two different techniques of classification. It was for Jonas, and other GIS technicians and analysts who invited me to join them for fieldwork, a social and sensual process through which they aimed to take on a role that went beyond simply representing a landscape and toward engaging socially in its production. He was merely one example of cases in which conservation professionals described to me the ways GIS tools became a means by which not merely technical, but far more personally and socially emotional anxieties and uncertainties, were worked through. Ground truthing then served two conflicting ends. It was a method for these conservation professionals to be *in* what they considered nature, to experience it and connect with it in ways that were fundamentally different than the GIS images of their everyday work. It was a desire that seemed to suggest a perceived shortcoming in the representative capacity of GIS. But, by justifying these more sensual explorations of nature through a technical need to ground truth, they were in the process paradoxically producing streams of validating data that acted only to strengthen GIS's technical claim to being able to perform tremendous representational work for nature.

Jonas had learned, through expensive graduate training, to process nature through GIS because it was a form of environmental labor that was in tremendous demand in the conservation labor market. Skilled labor in GIS is desired because the technology is powerful for producing the abstracted representations of nature that have quickly become dominant in contemporary conservation. Through these digitalized representations of complex landscapes, it becomes possible to create more universal technologies for valuing a fungible nature. Value becomes something identified from above rather than produced relationally within these landscapes. This ability to remotely render images and valuations of nature is the apotheosis of what Ingold has called the "the global outlook," the notion that through technology we can imagine and act on the world from outside of it, looking down on the globe as a whole environment to be acted on.[65] Ingold argues that this "movement from spherical to global imagery is also one in which 'the world,' as we are taught it exists, is drawn ever further from the matrix of our lived experience." He continues, "It appears that the world as it really exists can only be witnessed by leaving it, and indeed much scientific energy and considerable resources have been devoted to turning such an imaginative flight into an achieved actuality."[66]

"There's no 'ultimate truth' in classification," Jonas said to me as we wrapped up a day of work in the office later that week. You can only hope to improve the accuracy, and you improve the accuracy "with more and more people working in the area."

"Like the Mennonites?" I jokingly asked.

"No, no, no, not that kind of work," Jonas responded with a low chuckle.

But you really need more than just more people working, he continues, "a lack of ground truthing, *real on the ground* ground truthing, is our biggest deficit," he continued. You could see it in his eyes right then. Jonas was already back in "the field"—as he did some final corrections to the classification colors in thirty-meter-square pixels one by one, his distant gaze suggesting he was a maker of images, imagining himself in each pixel.

# 5

# THE POLITICAL ECOLOGY OF THE SEAWALL

## BUILDING THE LINE BETWEEN LAND AND SEA

Making the mapped boundary between land and sea appear real is a wet, muddy, low-paid job in Belize. Sean and "Rebound" were standing shirtless in about three feet of now turbid water. The depressed waistline of their soaked, cutoff jean shorts barely showed above the brown water. They each had a heavy shovel with a wooden handle and a rusted metal blade. It was the middle of the day, and the sun was shining straight down on the uneven surface of the sea, reflective sparkles creating a multifaceted sensory assault. They used the shovels to dig sand from the seafloor about twenty yards out from the shoreline of the property of the American lawyer that they were working for on a technically illegal project. With great effort, and low return, they would lift sopping wet shovelfuls of densely heavy sand up and into the well of a small dugout canoe. It was the type of canoe traditionally used by San Pedranos in small-scale lobstering operations but was used here instead to briefly store the dug-up remains of the lobster's nearshore habitat. When the small canoe reached its capacity of sand, they walked the canoe back to the shore of Mr. Sullivan's house, beached the vessel, and then unloaded the sand into sagging wet piles behind a crudely constructed log retaining wall built at the water's edge. They did this for about eight to ten hours (with ample breaks) each day for two weeks. Mr. Sullivan was paying them each BZ$50 (US$25) per day for their work.[1]

After they were finished the job, the line between the Sullivan property and the sea, between people and nature, seemed clear (though the water itself remained turbid for some time). The property did not begin, however, at the newly constructed wall. The property line, imagined as fixed but in reality as fluid as the sea that it abutted, was actually deemed to begin exactly sixty-six feet inland from the high tide line, now defined as where the wall of sand emerged from the sea. Those first sixty-six feet are the "Queen's Beach," a Belizean colloquialism for the linear commons along the shoreline intended to ensure public access to the shore and water. This coastal commons is an extension of British common law. But to "defend" his land from the sea, Sullivan first needed to *armor* the Queen's Beach. Technically, this structural intervention into the commons is illegal without first securing approval from the town and then a complicated national-level permitting process, but since the private property line always begins sixty-six feet behind the mean high tide line, beachfront property owners are left with limited means of intervention to stave off ongoing coastal erosion, a process that while to some degree "natural" has been exacerbated by shoreline development and the removal of nearshore seagrass (a problem to which the Sullivan property was, of course, now further contributing; figure 5.1).

I had been following their work curiously for those two weeks because the wall they were building was "armoring" the closest public shoreline to my family's temporary home, which was set

FIGURE 5.1 Collapsing seawall. Ambergris Caye, Belize.

*Source*: Author

back two lots from the Sullivan property (there were only about four lots between the sea and the rear lagoon at this narrow location on the island). Following several decades of tourism-driven gentrification in San Pedro, American and Canadian migrants now own most of the seafront lots on Ambergris Caye.[2] Sullivan, a white American, was a seasonal resident and transplant from the Pacific Northwest of the United States. I was invited at the completion of their work to drink rum with Sean, Rebound (a nickname that carried over from his time playing basketball in Belize City), and Sullivan on the high, wide veranda of his charming but architecturally incongruent, Southern plantation–style cottage.[3] The workers joked with Sullivan that this was good work because it paid well and came with free rum drinks, and, most importantly, they'd "get paid to come back and do it all over again."

Sullivan joked, "Oh fuck no," cursing freely as white expatriates are wont to do in a somewhat awkward attempt to bridge the age, language, and socioeconomic divide. "You guys have better done a decent job—I don't want my yard falling right back in with the next hurricane."

Sean looked at Sullivan, smiling, and said, "But that's just how it is."

## SEAWALLS AS CLIMATE LEDGERS

Discussions of natural capital often center around questions of value and its ongoing production. It's a model for nature that reimagines ecological processes as sites of active value production, as relational ecologies that function like capital to produce value on an ongoing basis. They deal less frequently with questions of debt and loss. In this chapter, I use the seawall as an entry point to consider the material and symbolic process of erosion and the production of ecological debt under climate change. I think about processes of erosion, simultaneously geologic and social, and the engineering of seawalls to trace relationships of debt between people and the coastal landscape. As well, I consider the progressive potential role of natural capital–based conservation tools, despite their ideological origin, in accounting for and repairing relationships of debt that are bound up within the much larger structure of extraction and loss made historically through colonization and climate change.

In this way, seawalls, in their building and shifting position along the shore, act as a sort of ledger that materially marks the unstable effort to manage and maintain value on the changing shores of Belize. The eroding shoreline is representative of the way the water's edge has become a site of calculation for understanding the burgeoning debt created by global capitalism and its emissions. Shifting high tide lines, and the loss that comes with erosion, registers forms of debt on this climate ledger. The seawall is then an attempt, however untenable, at balancing the books and creating fixity for the value of eroding coastal land.

But the seawall is also an attempt to sure up the overlapping modes of powerful social construction of nature and culture that underly critical fictions of capitalist value, most notably its abstraction from both the land itself in the form of private property and from the racialized bodies of people such as Sean and Rebound. Yet, as Katherine Yusoff remarks, the shoreline or beach is also a site in which these fictions face constant threat of erosion and slippage. "At the beach," she says, "the seemingly robust reality of social construction begins to slip, denoting a littoral zone of planetary counterplay among the altered physical environment, the psychic space of temporal location, and its historicity."[4] It is as the Caribbean philosopher Édouard Glissant points out, a space in which the contested relations between order and chaos are played out.[5]

Similarly, Cara Daggett, in her discussion of ideological confluence of climate denialism, racism, and misogyny that she calls "petromasculinity," says that with the seawall and its failures, there is a sense of emerging limits to containment and control. She gestures to the fact that the perceived surplus spaces of nature have long been used to engage in creative accounting of nature's costs, but, she says, "the industrial, capitalist system built to hide its waste, with its subterranean pipes, shunted refuse, oceanic dumps and accounting fantasies, is breaking down. Death is overwhelming the seawalls and flooding back into the carefully sanitised cores of privilege."[6] Here, the seawall is both a material fact of strained infrastructure and a metaphorical object for illustrating the limits of a colonial capitalist system dependent on the fiction of neat Kantian divides between nature and culture, sea and land.

The tide lines, and the seawalls that hope to mark its fixing, serve then as a site for an inevitable final accounting in the ledger of credit and especially debt created between people and the environment under climate change. While a great deal has been written about seawalls both historically and in the context of their role in stabilizing imagined futures under climate change, my focus here is on the way that the mix of technocratic and vernacular seawalls that have emerged in coastal San Pedro gesture toward competing efforts to secure value that feels increasingly threatened by climate change. Following work that explores the social life of infrastructure,[7] and that accounts for the agency of "vibrant" matter in shaping the way that humans are imagined to relate to and come into being in the material world, this chapter draws on the seawall, and the labor of its making, as a site of infrastructural imagination.[8] The seawall becomes an object over which questions of time, value, and property on the shoreline are understood under the increasing threats of climate change.

If sea level rise is the indexical phenomenon of climate change, then the seawall has emerged as the comparable form of infrastructure. It is by no means a new technology. There is evidence of durable concrete Roman seawalls that are at least two thousand years old and likely far older examples for which there is not an enduring material record.[9] Then, as now, the seawall is imagined as a technology for mediating that fraught intersection of land and sea, controlling and containing the unpredictability of an often-tempestuous sea. But this technology has taken on new meaning in the context of global climate change. From large projects authorized by governments to less formal and more fleeting efforts, seawalls mark responses to and imaginations of sea level rise across the world.[10] And, more often than not, stories about seawalls are stories about failure, not triumph. As Sarah Vaughn points out, despite a great deal of scholarship on the social life of seawalls, relatively little attention has been paid to the way "erosion problematizes modernist divides between nature and culture."[11] They are in a sense performative objects for seeing the way global climate change erodes any remaining sensibility of a neatly maintained nature-culture divide.

My ethnographic attention in this chapter is focused more on what might be called folk engineering or vernacular architecture.[12] That is, the work to reimagine the boundary between land and sea here—through built walls, mapping, and rhetoric—is generally not the project of a powerful state (or the performance of an aspirational state power) and large cadres of planning officials and architects. It is instead smaller in scale and reach. In some ways, this reflects a longer history of fractured, hesitant state intervention into the Belizean landscape. Much of the country, including its original capital city, was made through processes of reclamation and wall building. Reclamation work in Belize City began as early as the 1800s in an effort to make the colony's largest settlement more stable and permanent.[13] Mangrove swamps extend inland up to fifty miles, making the entire coastline porous and ambiguously delineated, a fact that both brought the country into being (it was difficult for Spanish colonial officials to enforce their claims) and persistently called into question its ontological status as a place and state (it was squishy, wet, and shifting). Despite this long history of intervention into the shoreline, in coastal Belize seawalls are typically small, fragmentary, informal, privately funded, materially suspect, and more of a collective expression of anxiety and hope than a centralized project of projecting futures. As such, they are sites of constant erosion.

## CHALLENGE TO VALUE AND COMMONS

Sullivan's seawall, and its limitations, point toward an underlying tension that lingers alongside new efforts to value intact nature in economic terms. The same nature that ecologists and economists seek to assign capitalist value for simply being is constantly having its very being threatened by a changing environment. This prompts interventions in the name of protecting or, as we shall see, producing private property in a manner that ends up undermining the functionality, and thus value, of the larger ecosystem. This illegal and imperfect intervention occurs at the revealing intersection of the imagined fixity of private property lines and the always-shifting liminality of the shoreline. As Bonnie McCay has said, "If property depends on clearly defined boundaries and some continuity in the thing itself, then coastal/marine property is complex and problematic indeed" because "littoral boundaries are liminal; they are complex, contextually contingent, and changing, challenging those who want solid and easily defined property rights or management jurisdictions."[14]

The littoral zone, typically defined as the tumultuous area between the high and low tide lines, is also where the lines between subject and object, land and sea, loss and gain, payment and debt, and destruction and creation all blur. It is a site of great fights for control—political, legal, economic, and material—because it is a space that challenges the very binary relationships by which control, power, and knowledge are imagined to be exercised in relationships with nature.[15] It is unstable and unrelentingly processual. The upland limit of the littoral zone, the mean high tide line, is the fiction on which the property line also typically rests. In some jurisdictions, the littoral zone is the commons, and private property begins at the high tide line. In Belize, the private property line is considered sixty-six feet inland from that line. Thus, it is through the littoral zone that we draw the boundary between the space of the imagined Lockean man whose presence and dominion is determined through his transformative capacity on land and the imagined "anticivilization" of the ocean.[16] In a nod to the insubordinate littoral territory, the state frequently intervenes, as it does in Belize, to mark this space as neither man's nor nature's, a space in which we all have the privilege of considering the very way this relationship comes to be constituted, a social space.

In this chapter, we will move from Sullivan's seawall to erosive hurricanes to considerations of ecological debt in order to examine the way the shoreline, and infrastructural efforts to secure its boundary, comes to function as both a theory machine for the social scientist and as an accounting mechanism of sorts for the public that seeks to narrate the meaning and effect of climate change on a rapidly changing island. The chapter then reads across the littoral zone to consider the way infrastructural interventions in the coastal zone point toward the ecologies of inequality that define the Anthropocene and the modes of accounting that seek to make this visible. As a place of instability and cyclical disaster, the tropical coastal zone is a place of anxious anticipation. And the seawall, and people's attempt to narrate its meaning, then becomes an object through which narratives of anticipation, disaster, and fleeting value come to be told in vivid ways.

The seawall, simultaneously a form of construction and destruction (in its removal of habitat), is an object through which anxious compromises about the relationship between people and nature, land and sea, and the present and future are made in material and narrative form. Seawalls were not central to my initial inquiry while working in Belize, but with time my ethnographic work revealed that people in Ambergris Caye had strikingly passionate feelings about these often-modest retention walls. They were frequently talked about with disgust for their aesthetic and presumed ecological impacts, but they were also understood as an inevitable compromise in a coastal space with soaring property values and multiplying subdivided seafront parcels. The wall and these parcels fronted a sea that came to embody all of the narrative possibility of an uncertain climate future, and various attempts to engineer that relationship revealed the highly distributed nature of agency in the making and unmaking of coastal value.

The building of seawalls is frequently referred to as a "hardening" or "armoring" of the coast, suggesting that the wall affords a way to limit the power or shaping influence of the reckless, uncivilized sea. But, of course, as the many ecosystem service models of the Belizean coastal zone attest, the shoreline and the sea are made through each other. Attempts at drawing a line between them, imaging them as separate functional entities, over time destroys the very hardened vantage point from which we imagine observing and knowing the sea.

It seemed then a curious coincidence that at the very moment this coastal island and the offshore reef were being increasingly drawn into a calculus of value premised on notions of the fluid connectivity of ecosystem functions that many people were nonetheless armoring their seafront properties in an apparent rejection—material and philosophical—of these influential new narratives of value. The rejection of connectivity implied by the building of seawalls, I argue, is not simply about protecting a property line (though that is a part of it) but protecting and making manifest a line of influence and agency in the face of profound uncertainty.

Ironically, the modeling of ecosystem service value, made through the imagination and simulation of loss and disaster, feeds into the very angst that drives the building of seawalls—the sense that it could all be lost. The effort to safeguard ecosystem function in the natural capital approach is described as a form of social insurance against disaster scenarios, but the seawall is a much more individual form of insurance. The progressive thread of the natural capital approach is that it compels planners and citizens to consider their future socially, but it does so through a calculative language made from managing private modes of ownership.

The language of ownership in environmental projects, Alex Golub has pointed out in Papua New Guinea, is more the "the internalization of a certain emotional disposition."[17] His definition highlights the fact that the language of property and finance that drives natural capital and ecosystem service conservation often ends up being more about achieving certain affective changes—reorientations of social relations—than it is about creating a currency of value that resides somehow outside of those relations. This affective notion of ownership is often in conflict with forms of private property ownership as a set of legal rights.

Some of the surprising impacts of natural capital discourse in Belize—a professed sense of anxiety on the part of citizens engaged with the idea and the lack of an actual "market" in these services—can be traced in part to the tension between the desire to enact a social and ecological future through tools and narratives that still allude to, and perhaps hope for, a presumed calculative individual. My fieldwork with ecosystem service scientists suggests that in many ways their approach to the project of conservation is born of a genuine desire to promote more relational notions of the ecological landscape and its value. While it is certainly not an entirely postmodern obliteration of the subject-object relationship central to Western notions of nature, their thinking does seek to consider the way life for both humans and nature is deeply co-constituted and how in a world that challenges the very divides upon which management and governance have been premised, we are forced almost naturally to a more ecological and social understanding for considering the future.

In its earlier use as a mere rhetorical tool, that was the very point of the natural capital concept—to point out that nature and people are inseparable, and thus value for humans cannot be made without sustaining nature. When the conservation scientists I spoke with used the word "inseparable" to describe that relationship, however, they were generally imagining two discrete entities in a close embrace, not the breakdown of the very categories that produce these entities. A natural capital approach also then has the capacity to challenge romantic Lockean notions of private ownership embodied by the seawall. Rather than working from the premise that value and ownership are established by transforming natural processes into managed human processes, under the natural capital model, effective or even transformative stewardship comes often in the form of *not* intervening materially. These new forms of human-made value in nature can only be realized by *not* actively transforming the landscape. It is a potentially disruptive take on the production of and responsibilities under regimes of private ownership in the coastal zone.

In this context, seawalls then anticipate loss and disaster. They defend against that disaster for individual property owners (for a time), but for society more broadly they add to the ecological debt that storms represent (as the possible outcome of global warming) and make that evident (through their uneven effects) by breaking the connections that sustain ecosystems. The building of seawalls is an attempt to sustain the illusion of a Kantian divide and the kinds of value and power relationships such a divide makes possible. Their frequent failure, along with the positive feedback loop of detrimental ecosystem effects that they unleash, point to the limitations of the narratives of control, ownership, and value on which they are built. The efforts to manage the anxieties they represent legally, socially, and ecologically on Ambergris Caye point toward the spaces of possibility in an ecosystem service approach, as well as the rhetorical shortcomings of such an approach: It relies on a language that refers to an entirely different relational ethics of value than that which it actually aims to instantiate.

## SEAWALLS AND THE RETENTION OF VALUE

The littoral and liminal space of the coastal zone troubles conventional spatial representations of value in market-oriented conservation projects. The coast is shifting and impermanent. It is not in reality, a line, polygon, or point (the basic means by which a feature can be represented in GIS). It is a relationship. The result is conflicting but overlapping experiences of and representations of the coastal zone.

McCay uses the concept of liminality to describe the uniquely challenging management features of the coastal zone:

> In cultural theory, liminal signifies phenomena that do not fit well into cultural categories; the ambiguous, the ambivalent, the untidy, and very often therefore that which is treated culturally as sources of fear, risk, and danger, or simply forbidden (for example:

> Firth and Leach 1967; Leach 1964). Coasts are valued aesthetically and for their many critical functions, economic, social, and ecological. But they are also liminal, in-between and both land and sea, bathed by both tides and "run-off," sculpted by forces of wind and waves that intersect and play upon the consequences of plate tectonics and bulldozers. Their boundaries are often vague, shifting, and contested. If not feared or tabooed, they do pose cultural, political, and survival challenges and opportunities.[18]

The coastal zone is an area of intense interest and intervention but a space in which planned intervention and investment is constantly undermined. In the coastal zone, attempts to formalize and standardize the relationship of value between land and sea, and people and the physical environment, manifest themselves in a complicated web of legal and social conventions on use. The attempt to resolve the tensions of these overlapping conventions of management and use is in some ways an effort to fix forms of value and techniques of valuation that are made historically in the context of more stable and predictable landscapes. Ecologists argue that the ecological value of the coastal zone is in the fluid relationship of exchange between organisms and the land and the sea (and regular people who enjoy the sounds of crashing surf; the twice-daily revelations of changing tides are valuing a similar form of exchange). But, in the short term, attempts at fixing this value for an individual property owner often depend on a disruption of those relationships by creating a stable line between land and sea. It is here at the very edge of its existence that land takes on its greatest value but also where the relationships and institutions that traditionally sustain the complex fiction of valuable land are most prone to disruption (figure 5.2).[19]

It is in these spaces of disruption, Heidegger argues, that perhaps paradoxically the subject-object relationship comes

FIGURE 5.2 Aerial view of the Mahogany Bay Resort under construction in former wetlands.

*Source*: Author

into being. He doesn't write against the notion of a subject/object dichotomy but "points out that our experience of this subject/object relation *derives from* and so *presupposes* a more fundamental level of experience, a primordial modality of engaged existence in which self and world are united rather than divided."[20] When our practical engagement with the object world breaks down in times of crisis, disruption, or confusion, people suddenly come to experience themselves "as isolated subjects standing reflectively before a world of external objects, which we thereby come to experience as standing over against us in the mode of something objectively 'on hand' (*vorhanden*)."[21] The coastal zone is a site of continuous disruption to the ordinary engagement with the object world. For this reason, it is an incredibly useful space to think about the relationship between what is thought of as society and nature, and it is a site of frequent harried human interventions seeking to simplify and order the crisis of disruption.

Like the crass financial calculation of land and value that it seeks to make possible, the seawall attempts to end the messy ongoing relationship of exchange. The seawall is an attempt to make real a subject-object relationship, a relational fixity of value that is in the end impossible to sustain. "The world," Marshall Sahlins reminds us, "is under no obligation to conform to the logic by which some people conceive of it."[22]

In the end, the world is unlikely to conform to Sullivan's modernist vision. And the work of Sean and Rebound on the Sullivan property will probably not last long, certainly not in geologic terms. The modern (geologically speaking) Ambergris Caye formed approximately seven thousand years ago. The prolonged period of sea level rise following the peak of glaciation eighteen thousand years ago allowed for a new reef to grow and for the island to form as an accumulation of carbonate sand and mud that rested on top of an older Pleistocene limestone formation. The lumpy, shifting mass they were augmenting with their work is the latest iteration of Ambergris Caye in a nearly two-hundred-million-year period of alternating inundation and emergence for the broader Yucatán Platform of which the caye is a part.[23]

During these periods in which Ambergris Caye persists as something we might call an island, its shape is radically and regularly altered by the force of hurricanes, but they do not destroy it. Rather, they reshape it, resetting the terms of engagement between people and the material world of the island. Like in many coastal areas, in an era known as the Anthropocene, the boundaries between the forces of influence of people and nature become ever more tenuous in coastal Belize. Warmer waters, it is argued, lead to more frequent hurricanes, while more acidic waters contribute to the degradation of the sheltering reefs. People are literally remaking the very nature they stand against in a more nefarious form.

Thinking about these disruptive changes as the outcome of anthropogenic climate change also leads to a blurring of temporal scales. Climate change is a long-term disruption that comes to be a thing through the statistical coalescence of a series of sometimes traumatic, immediate events. The hurricane itself is neither natural nor human. It is both. The relationships that these storms—and this climate—create at the shoreline have no dominant party. And the relationships of value made in this complexly illustrative space are ever more critical but increasingly difficult to model and distribute in ways that ensure equity and stability for an unevenly vulnerable coastal society.

Despite fear of the violence of climate change, it's been over two decades since a major hurricane struck Ambergris Caye. The next time a hurricane hits the island, the boundary between the Sullivan property and the sea will be violently blurred. The carefully arranged sand that formed the seawall will be redistributed with tumultuous disregard for survey maps or models of coastal ecosystem function. Some will find its way to the sand ridge that forms the wiggling spine of the island while some sand will find its way back to the seafloor from which it came.

## THE POLITICS AND POSSIBILITY OF CLIMATE DEBT

The last major hurricane to seriously rearrange the eastern, seafacing side of Ambergris Caye was Hurricane Mitch. Although northern Belize suffered only a glancing blow from the monstrous storm, the massive swells destroyed docks and caused severe beach erosion throughout the country. Further south in neighboring Honduras, at the southern end of the Mesoamerican Reef, Mitch was one of the most devastating Caribbean storms in recorded history. It stalled over the high terrain of Honduras resulting in a scale and intensity of damage so severe that it gave traction to entirely new ways of thinking about the relationship between natural disasters, political organization, and value. Most notably the devastation led to one of the first widespread applications of the concept of "ecological debt." According to Erik Paredis et al.,

> The ecological debt of a country A consists of (1) the ecological damage caused over time by the country A in other countries or in areas under the jurisdiction of other countries through its production and consumption patterns, and/or (2) the ecological damage caused over time by country A to ecosystems beyond national jurisdiction through its production and consumption patterns, and/or (3) the exploitation or use of ecosystems and ecosystem goods and services by country A at the expense of equitable rights to those ecosystem and ecosystem goods and services of other countries, or individual.[24]

The discourse of ecological debt first developed in the South American environment and development NGO community in the early 1990s. While it was a powerful rhetorical tool, economically dominant countries did not line up to repay their debts. However, in 1998, following the devastation of Hurricane Mitch, several countries, including Honduras and Nicaragua, appealed for debt forgiveness based on the notion of ecological debt, arguing that the decade's striking increase in storm frequency and intensity in the Caribbean Basin was at least partly attributable to climate changes caused by the world's industrial economies.[25] If the economically dominant countries would not pay the ecological debt owed, the ravaged countries argued, they could at least forgive the financial debt owed by the poorer countries. Most nations (except for Cuba and France) and the World Bank, however, agreed only to suspend payments for three years. However, interest would continue to accumulate on the debts during that time.[26] The concept has occasionally stirred to life from time to time in intervening years, especially more recently in Africa.[27]

Interestingly, economists such as Richard Norgaard, who has been critical of attempts to operationalize ecosystem services as actual monetary values, have been less critical of the inverse project—calculating ecological debt—and less reflective about the way these two forms of calculation are intimately related.[28] As Sigma Colón points out, "How nature's resources are used or preserved is determined by the twin processes of measuring and valuing, which are intimately tied to notions of debt."[29] In fact, in the case of Paredis's operational definition of ecological debt, you can't calculate the uneven distribution of ecosystem goods and services until you first have some baseline notion of what an ecosystem service is and how one might value it.

Colón, in discussing it, highlights the more nuanced engagement by critics of monetized nature when she says, "For developing countries in debt the earth's resources provide a tangible and immediate form of capital."[30] Members of the Natural Capital Project, and other similar endeavors, would likely agree and argue that their continued economic wealth and social stability is dependent on a more thoughtful management of that natural capital. In a similar vein, literary critics such as Margaret Atwood, as Colón points out, have even gone so far as to argue that nature itself "is an expert in cost-benefit analysis" though "she calculates" in a different way than humans.[31] Nature doesn't request repayment; she demands it through disaster and plague, resetting the books violently.

There are two things worth noting in these discussions of ecological debt. First, they emphasize how intimately notions of debt and destruction relate to ideas of value and reconstruction. Second, they highlight the inconsistent and at times contradictory engagement of progressive critics with narratives of nature's value. There is little hesitancy to calculate in precise dollar terms the value that is taken away through the socio-ecological exploitation by industrialized economies but a great deal more reticence when it comes to calculating the potential ongoing productive monetary value of that capital when it *isn't* exploited or destroyed.

There is a willingness to calculate destruction but not production, which raises the obvious question of why? And how does the failure to treat the sides of the value equation equally divert social scientific analysis from understanding more fully the intimate nature of their co-constitution? I began my work in Belize in 2009 because natural capital was an emergent language for talking about the value of nature, especially along the Mesoamerican Reef. Also in that year, the climate ministers of the Central American nations met in neighboring Guatemala to discuss the ecological debt owed by the industrialized countries to Central America. Their meeting was little covered in the global media, but in a hastily and crudely derived figure (and probably a gross underestimate by most better measures), the ministers declared an ecological debt of $105 billion. They cited the region's heightened vulnerability to climate change through increased hurricanes, prolonged droughts, freak storms and tornados, and other rather frightening problems. I came across an article about the meeting in the *Amandala*, Belize's left-leaning newspaper. The article said that the ministers "chose to put out the preliminary figure to drive the point home about the dire costs of climate change being caused by industrialization."[32] It was the last time I heard the number mentioned. Over the next five years, the *positive* value of the reef as capital caused much more anxious concern.

How does a calculation of a debt owed because of the deadly present and future under climate change cause less anxiety than the idea of an immensely valuable form of naturally occurring economic capital simply waiting to be exploited? As a form of explicit and corrective claim-making, many people I spoke to found the narrative of debt to be empowering. As an object made through existing relationships of value, through largely predetermined arrangements between people and landscape, the idea of natural capital was a form of value that existed within and through an already uneven social landscape (in some sense natural capital and the language of responsibility is a rehashing of the traditionally uneven relations of moral power around debt). This new form of value for most people was too intimately tied to old relationships of value. As Marx noted, capital is relational and processual—not simply something waiting to be seized—and the relationships that make the imagination of this form of value possible were often ones of ingrained anxiety for the Belizeans I worked with. That is how, despite the empowering (and frankly often appealing) rhetoric of value produced under the natural capital model, many of the most marginalized people in the country, and likely elsewhere, may feel more empowered by the capacity to claim a right to a debt owed by articulating new relationships and explicitly monetized mechanisms of payment than by far more vague notions of value made through shifting and abstracted historic social and political relations of an uneven landscape.[33]

Hurricane Mitch followed an unusual course. It originated in the waters just north of Colombia then moved north before making a sharp westerly turn just south of Jamaica. It appeared to be headed directly for Belize, but the storm then made a last-minute turn to the south toward Honduras, stalling and spilling its energy until exhaustion in the higher terrain of that country. The late turn spared Belize much of its direct wrath, but three years later, northern Belize was hit by another powerful and unpredictable hurricane—Keith. Like Mitch, Keith stalled just as it made landfall. The storm spun erratically in the area of Ambergris Caye for three days, creating a persistent northerly wind that people say blew the water out of Chetumal Bay on the leeward side of the island, prompting people to walk out onto the empty bay floor from the mainland in a Moses-like moment.[34]

For residents of Ambergris Caye, it was less Exodus and more Genesis as that water pushed up and over the island and resulted in three days of flooding. The prolonged storm surge from the west combined with torrential rains (32.67 inches were measured in Belize City) resulted in serious damage on the island, but it was distributed unevenly. The windward side of the island, where most of the tourist infrastructure is located, had surprisingly little damage (it fared worse in Mitch). Even the often-rickety wooden piers mostly survived, and while a few buildings lost roofs, there was virtually no destruction there. However, on the leeward side, over fifty homes were destroyed and hundreds of others damaged. Boats, typically tied up in the lagoon side to ride out storms, were blown ashore and destroyed.[35] Keith and its wandering path were partly to blame. The persistent northerly winds pushed the bay waters onto the rear of the island (which runs southwest to northeast at a roughly thirty-degree angle) with unrelenting force.

But the extensive damage was also the result of processes of geology and political economy. The limestone island was made through the accumulation of reef debris that washed back over the uprising that became the island along with the calcified remains of shellfish that lived in those initially muddy recesses of debris over thousands of years. The result is that most of the highest land is near the eastern (front) shore. The land then sags and is often swampy toward the western lagoon shore. The highest land was most desirable because it was the driest and most stable and had the added benefit of cooling and mosquito-diffusing breezes. The highest land on Ambergris Caye, just inland from the eastern shore, is sometimes referred to as a "sand ridge." Poorer villagers typically lived on the lagoon side of the village. During the 1980s and 1990s, the island's typical "West Indies"–style wood-frame cottages were increasingly replaced by cement block structures. This transition generally followed the contours of wealth, so by the time Hurricane Keith struck in 2001, many of the front-facing buildings were sturdy cement-block construction on higher land while the lagoon-side buildings were still largely wood-framed ones on low-lying reclaimed wetland.

The unusual course of Hurricane Keith dramatically highlighted this uneven topography of land and wealth. It was the unevenness of the distribution of environmental losses that is made so explicitly clear by natural disasters that brought the idea of ecological debt to the fore in the 1990s. Natural disasters, they argued, were social and political products, the outcomes of forms of vulnerability made through explicitly social and political processes.[36] The movement emphasized the uneven global distribution of vulnerability that came about in large part due to the uneven distribution of the wealth of nature's resources. Keith, like so many other disasters, shows the way that this unevenness plays out at a variety of scales, including the local.[37] And it underlines the point that discussion of environmental value shouldn't be simplified into a debate about debt versus wealth. A truly social and productive science for valuing nature must confront both debt and assets, production and destruction. It is a science that must come to terms with its narrative origins in the imagination of disaster and in the practice of theft.

## CLAIMING VALUE IN DEBT AND PROFIT: THE PEOPLE'S REFERENDUM

In winter 2012, with growing pressure from international business interests and some members of government to exploit Belize's potential offshore oil resources, a group of environmental NGOs in the country launched a signature drive to initiate a nationwide referendum on offshore oil exploration. The group collected thousands more signatures than required, but in a highly disputed certification process over eight thousand of the signatures were declared invalid. As a result, no state-sanctioned referendum would be held. Angered but unwavering, the coalition began organizing an unprecedented event—"The People's Referendum." Partnering with justices of the peace around the country, they set up mock polling centers for each town in Belize and held a nationwide referendum. Utilizing their own infrastructure of buildings, communications, and supportive justices, they produced a powerfully symbolic referendum vote in which nearly a quarter of eligible voters turned out, with over 96 percent expressing that they did not want to see offshore oil exploration in Belize.

It was a remarkable event for several reasons but especially as a striking demonstration of the evolving location of conservation politics and practice in Belize. The coalition of environmental NGOs that organized the symbolic referendum had in the months beforehand distributed materials indicating the economic and social value of the reef, frequently expressing that value explicitly through the language of ecosystem services. The tagline for the campaign was "Too Much to Lose!" and that loss was made in direct reference to the "over BZ$1 billion annually in goods and services" contributed by the reef. The language was widely adopted by citizens in conversations about oil drilling and the discourse about oil drilling came to be a debate that pitched an extractive resource whose benefits would be distributed narrowly and potential losses widely versus the services of the reef (tourism, fisheries, storm protection, etc.) that broadly support the Belizean people rather than a few self-interested businesspeople and politicians.

The language of ecosystem services, and the debt associated with their potential loss, became a potent tool for expressing nature's value in a way that simultaneously challenged the claims of the Belizean ruling governmental party and those of the largely foreign oil business interests. When the government failed to respond to the growing chorus of concerns, the coalition of NGOs and citizens proclaimed their own governing authority that was founded in a remarkably wide-reaching conviction in the value of the Mesoamerican Reef. The referendum launched an ongoing opposition campaign that eventually led, in 2013, to a Supreme Court ruling that canceled the oil contracts previously granted by the government.

There is then a great deal of potential political power, and a power that is distributed along new lines, if these forms of environmental value can be articulated in terms of a right not only to a "just share" of a nation or community's wealth but also to a broader participation in the politics of the distribution of those rights. The narrative of ecosystems services was effective in the case of the People's Referendum because it was a means toward offering an explicit critique of the existing relations by which economic value is made and distributed in Belize. It used the specter of a societal debt imposed by wealthy outsiders to mobilize new societal claims on existing forms of value. Activists in Belize then used the tools of ecosystem services, a framework most widely introduced by the Coastal Zone Management Authority and Institute (CMZAI) through its collaboration with the Natural Capital Project. The CZMAI had been reauthorized (brought back into being) by the current government, and its tools for valuing nature were then used to question other decisions made by that same government, namely offshore drilling concessions. An ostensibly neoliberal interpretive grid for valuing the natural world was used to undermine a government plan to rapidly privatize off-shore oil resources for sale to second-rate American oil exploration firms. It drives home James Ferguson's argument that a surprisingly "radical political indeterminacy . . . attends any specific governmental technique" and that great analytical insight and political possibility are to be drawn from avoiding the conflation of governmental or political technique and political philosophy.[38]

A concern raised by economists and conservationists interested in pricing ecosystem services is that the demand curve for those services, which are derived from nonsubstitutable natural capital, though difficult to truly model, is unlike that of "normal goods." Demand, and theoretically price, would approach infinity as the quantity or supply of an essential ecosystem service declines and approaches zero, up until the point at which the threshold of system collapse is reached and marginal value is no longer a relevant mode of analysis.[39] It could theoretically be more valuable then to hold a share or stake in an essential ecosystem service under times of great ecological disruption and threat to that service. This is true with most essential goods, but with forms of natural capital such as water, conservationists argue, we can't move to a replacement good; we die. With a market in these services, it could theoretically be profitable for an individual with a short enough time horizon to invest in marine ecosystem services and offshore drilling simultaneously.

One means of addressing the implications of this concern—that value in ecosystem services becomes simply another unevenly held commodity in a system that thrives on directed disruption—is to think more broadly about relations of value in nature by incorporating narratives of debt (monetary and nonmonetary) along with those of value and considering, and even calculating, the way ecological debt and ecological value are made together. There is a need then to consider the means for a more equitable distribution of existing ecological wealth and to consider more thoroughly how to redistribute the burden of ecological, and hence social, wealth already destroyed. It builds upon what Patrick Bond has called a "rights-based approach," which takes on an affirmative role in the north-south redistribution of both wealth and debt that have been distributed in radically uneven ways. Paradoxically, such a rights-based approach potentially requires using the tools of calculation developed through market capitalism and the expanded domain of those calculative tools seen under neoliberalism. A more holistic consideration of ecological wealth that uses the interpretive tools of the social sciences and does not shy away from the analytical tools of the market may permit a fuller consideration of how we organize society and the political tools and strategies by which we might imagine more equitable, inspiring environments in the future.

The world's oceans are the quintessential commons, largely outside of regimes of private property and other modes of state regulation. They push back constantly at the social fictions told more comfortably on land. The great paradox of this moment that some have referred to as the "Capitalocene" is that one of its most remarked upon material outcomes has been the rising of the seas and thus the expansion of this commons into what emerged in late capitalism as the most highly valued parcels of private property, the shoreline.[40] Rising sea levels then force us to confront the spaces in which the relationship between private and public, and commoditized and common, are made and contested. And in intensely developed and capitalized spaces like cities and tourist centers, this moment provides a forced consideration of the competing models of value that interface in these spaces. The seawall, in the end, seeks to mark and maintain a clear boundary between not just nature and culture but also between that which is private and that which is held in common. The sea, the quintessential messy common, is always engaging in a geophysical game of cat and mouse with the imagination of private property. In this chapter, I have emphasized the geologic agency of the sea and its capacity for materially and ideologically eroding the powerful social constructions that seek to maintain racialized capitalism in Belize and elsewhere. But in the concluding chapter, I turn to the ecological and biological components of this relationship with the sea. How does the sea not just reshape the shore but bring the people that reside there into new, sometimes undesirable, relations of commoning?

# POSTSCRIPT

## Weedy Coastal Commons

### ENCOUNTERING THE WEEDY COMMONS

Sitting on the beach during a prolonged sargassum inundation is a sensorial experience akin to sharing a confined office space with someone that enjoys hard-boiled eggs as a snack. Particularly when the onshore breeze fades, the stench of decomposing sargassum on the granular sand beaches and nearshore waters of Ambergris Caye has an inescapable depth to it. There is a sulfurous pungency to this greenish-brown, free-floating seaweed that, despite the cultural specificity of sensory experience, almost universally triggers alarm. Its presence was so alarming in spring 2015 that London's *Daily Mail*—dreading the ruined holidays of upper-middle-class colonial nostalgic Brits—gave the massive lumps of seaweed a violent agency and proclaimed that the weedy mats were "terrorizing Caribbean beaches."[1]

It is not a new component of the ecology of the greater Caribbean Basin, but in the past decade, it has washed ashore in Belize and throughout the Caribbean region in greater volumes and with greater frequency.[2] This seaweed covers beaches, clogs boat engines, suffocates small nearshore fish and seagrass, and turns clear waters a turbid brown. Despite its popular portrayal as generally undesirable, however, I argue that this newly prolific sargassum is a peculiar brand of weedy Anthropocene commons, an object of dreaded excess and potential opportunity that brings into stark relief the challenge and possibility of living as commoners in a time of profound climactic change. Unlike traditional imaginations of the commons, however, its presence sparks anxiety not about how we will judiciously distribute a broadly desired resource but how we can manage the distribution of something held in common yet wanted initially by nearly no one. Like so much of the unruly, emergent Anthropocene environment, access is not excludable, even if many wish otherwise. The prolific and often chaotic ecologies that emerge out of the excesses of disruptive capitalism, rather than merely foreclosing future possibilities, demand new ones. Sargassum is a weed that arrives ashore as "an excess or exudation that is prior to and product of capitalist accumulation that capital . . . can never fully capture."[3]

As such, it is an ecological disruption that demands new possibilities even as it seems to foreclose others. This disruptive weed is emblematic of a "permanently polluted world" in which the politics of response necessarily emphasize creative incorporation over conventional responses of "separation, containment, clean up and immunization" premised on an imagined "politics of material purity."[4] While the ecological excesses of capitalism often disproportionately impact the lives and livelihoods of marginalized people, sargassum's arrival has often been uniquely disruptive to the established political and economic elite while providing an everyday politics of possibility as it comes to be used or circulated by those on the social periphery of coastal Belize.

A stinky, unwanted, lumpish aquatic weed may seem a strange object around which to conclude a book about environmental value and its production under the crisis of climate change. But, in this chapter, I aim to argue that any understanding of the material world and its value in the Anthropocene must attend not just to social processes of abstraction and quantification but also toward an entangled relationality, a nature in value that is weedy. While a good deal of recent work has thought about weedy ecologies in terrestrial spaces, sargassum's unwelcomed arrival in the Caribbean Sea and Gulf of Mexico points to the way our lives are increasingly entangled in the weediness of a changing sea and the way the proliferation of weediness takes place not just through localized processes of disruption but temporally and spatially vast shifts to entire climactic systems.[5] In this chapter, I aim to expand what a relational idea of value might mean in a changing climate and, following Bird Rose, extend the realm of value beyond transaction and toward obligation and commonality that comes with being in the messy presence of others, human and nonhuman, with which we take on this time.[6] I bring together critical commons literature, including feminist approaches to commoning, with anthropological approaches to multispecies encounters in the Anthropocene to reveal the analytic and material productivity of sargassum in the Anthropocene Caribbean. Sargassum, I argue, should be understood as an illustrative brand of (un)commons that points toward the productive possibility of a quotidian politics of disruptive incorporation in an epoch that defies desires for the order, purity, and socioecological differences that have sustained the often exploitative social and material relations of capitalist ecologies of value. I make this argument ethnographically by following sargassum as it circulates across water and land and through the lives of Belizeans making a life at the margins of a bustling tourist center.

## IMAGINING A NEW COMMON GROUND

In early 2015, at the peak of the first major sargassum inundation (see figures Con.1 and Con.2), I was sitting on the beach one morning exploring this newly prolific seaweed with my five-year-old daughter and some other neighborhood children. Unbothered and curious, the children began picking up clumps of it, placing it on each other's heads, wearing it on their arms and shoulders, laughing hysterically. Their bodies shook with delight as they dangled the seaweed—simultaneously coarse and slimy, clumped and stringy, appealing and grotesque—over my head. Soon the sargassum slipped from the control of their tiny hands and landed atop my own head like a tentacled creature releasing tiny crabs and other macroinvertebrates onto my sunburned scalp. They declared in a joyful tone that seemed to make reference to a horror movie, "It's ALLIIIVE!"

And it was. Sargassum, while often portrayed in news accounts as a dead rotting singular object, is in fact a loose assemblage of wandering life. A host of small crabs and macroinvertebrates live within its strands. At various times, "rafts" of it serve as habitat, nursing ground, and refuge for "fishes, sea turtles, marine birds, crabs, shrimp, and more."[7] In fact, a few species live exclusively within the secure confines of these masses that the National Oceanic and Atmospheric Administration describes, fittingly, as "island-like." Sargassum serves as its own generative ecology amid the vast hostility of the open ocean.

The vitality of sargassum and its history as a dwelling place for vulnerable species are both striking. But once it reaches the shoreline, the seaweed begins to die and decompose, and it becomes in the popular imagination an object of dread. However, when we walked back to our house later that day, we passed a friend and neighbor, Roberto, who was transferring wheelbarrow

FIGURE CON.1 Sargassum piled along the beach. Ambergris Caye, Belize.
*Source*: Author

loads of it onto a handmade wooden trailer attached to a golf cart. The encounter sparked a conversation that helped me see the continuing circulation of sargassum as a lively object that was still generating new kinds of possibility.

Roberto did landscaping and maintenance work for a property near the beach in an expatriate enclave known as Tres Cocos. Like many young men in their twenties in San Pedro, he is not a native of the island community. He grew up on the mainland in the far west village of Benque, near the Guatemalan border. He married soon after arriving on the island. Young, married, and with a child on the way, he took advantage of what seemed like an ideal opportunity to buy a small piece of land for a home in the San Mateo subdivision. Unlike other large

FIGURE CON.2 Ambergris Caye beach prior to regular sargassum inundations, 2015.

*Source*: Author

subdivisions built in San Pedro in the last couple of decades of extreme growth, the land for San Mateo (and two others created in the late 1990s, San Juan and San Pedrito) was not purchased by the national government and prepared for development prior to its subdivision. Rather, it was subdivided by the town without an environmental impact assessment or any permitting through government agencies such as the forestry or fisheries departments. By the time this process was criticized, it was too late, and people like Roberto had pinned their family's small fortunes on the land they had been deeded.

But "land" in new San Pedro subdivisions remains more of an aspirational term. It describes spaces that have been marked and deeded—spaces that look quite hard and tractable on planning maps but in reality are more fluid and squishier. In some instances, this land is dense stands of mangroves with their roots clinging through shallow water to the muddy seabed. In even worse cases, it is wetlands with several feet of standing water, even at low tide. Flat-bottomed fly-fishing boats occasionally troll right across it looking for tarpon or permit fish. Where the San Mateo subdivision stands is a mix of mud and sand left behind by the shifting currents of a narrow, navigable inlet that links the front and back lagoon in San Pedro, a feature referred to as "the cut." There is one high sand ridge right along the current channel, but back from there the land is mostly mangroves and wetlands, which flood with each high tide. Nonetheless, in a rapidly growing, housing-starved island, the San Mateo neighborhood has grown tremendously over the last decade.

Homes there stand on wet wood beams anchored gingerly into the shifting ground. Without adequate money from the town or government to truly complete the San Mateo subdivision, residents have taken it upon themselves to "fill" the land, to make a place in which they can live together. In some cases, this means simply throwing garbage beneath the home and occasionally burning it to create a new layer of material. In other cases, people with access to the right equipment and social connections have dredged sand from the rear lagoon and filled their land with that material. This is often expensive and relatively capital intensive. But I was surprised at the time to learn from Roberto that the most common method for filling land in San

Mateo was with piles of seaweed and seagrass. He described taking piles of seaweed from the beachfront and loading them onto a makeshift trailer attached to his employer's worn-down bright red electric golf card and then "using it to fill [his] land, little by little" on the other side of the island.

Roberto explained the high cost of other kinds of fill and his reticence to simply use garbage to fill the land as some neighbors had previously. He hoped that with enough sargassum he could eventually fill his land. That idea ("filling land") has a contingent possibility built into the phrasing, the normalized assumption that making this marginal land emerge from water would require further entities and relationships. For Roberto, the sargassum represented the possibility of realizing his land and firming his position in this community.

He had been doing this for weeks, but it did not draw my attention at first because it was a fairly small-scale, individual project. But later, as the sargassum inundation continued, several men with access to pick-up trucks typically used as part of island delivery systems retooled their businesses to help the town government and private businesses remove the seaweed from their beachfronts. The San Mateo neighborhood quickly became the primary site for taking it because it was free to leave there, and residents were generally appreciative of the bountiful new fill material. Truckloads of it were soon being delivered daily to Roberto's home and others in San Mateo. Golf cart trailers quickly gave way to large pick-up trucks and then to the broad tires of large dump trucks that left deep irregular treads in the putty limestone streets of the neighborhood. Within months, seaweed covered Roberto's land. And as the seaweed broke down, slowly releasing its sand and itself becoming a mushy dirt, it began to literally produce Roberto's land—it was still wet, and still occasionally flooded, but out of water had come land.

As many people in San Pedro now say, confounding biologists and physicists, "seaweed becomes sand."

The seaweed that nobody wanted, and that still most don't like, Roberto told me the next year, "was like a strange gift. "The sea," he said, "always gives. . . . It's for all of us." Moving across the sea and through the labor of people in coastal Belize, breaking down and transforming, sargassum had made watery San Mateo into a delicate, if imperfect, refuge for a diverse assemblage of newcomers to this increasingly expensive, inhospitable island community.

## SARGASSUM AS *UN*COMMONS

Sargassum, I argue, functions as a type of "uncommons," an assemblage "of life where nature and humans might be beyond the demand for an either/or distinction," one that highlights the way people and objects like sargassum can come into common life together even as they may diverge from one another in interest and practice.[8] The idea of the uncommons, of being together with entities "while remaining distinct," captures something of the vulnerable assemblage that is the unsought after, often alarmingly chaotic, common life of the Anthropocene. July Hazard refers to these spaces that challenge the bodies—political, material, human, nonhuman—that conventionally constitute the commons as "a fracture commons—a commons constituted of toxicity, shattering, and trauma that occur at porous/permeable edges, as where water meets land."[9]

Building on Judith Butler, Irina Velicu and Gustavo García-López have written about this sense of fracture and vulnerability not as a pathology to be resolved through commoning but as a status fundamental to the human condition, which is "given by the many unchosen proximities that (re)-make us throughout our lives, challenging expectations about self-sufficiency or autonomous choice-making."[10] The push is to think of commoning as way of dwelling within and building from that vulnerability rather than seeking to overcome it through what de la Cadena calls the "disavowal of excess entities" that occurs when we seek sameness as a starting point for being commoners.[11]

These approaches challenge the history of patriarchal, heteronormative, racist, and classist formations of the commons and point toward the role of feminist and Indigenous ecologies in developing a stronger empirical basis for studying them.[12] They point out that institutionalist models of the commons often overlook forms of social differentiation to assume a community homogeneity of interests that acts to entrench existing privilege and inequality in resource management.[13] This absence of social difference is, Silvia Federici points out, especially important because society's most marginalized members often depend on access to an equitably managed commons.[14]

These relationships, the resources they are envisioned to protect, and the counterhegemonic worlds they aim to enable are therefore not simply waiting for recognition and care—they must be made and performed.[15] The notion of "commoning" used in critical commons scholarship is an explicit move beyond the commons as a priori objects to be managed and toward creating counterhegemonic relations of *being in common* with others—both human and nonhuman, and in a manner that meaningfully engages difference, including through and across species boundaries. The nonhuman, rather than an object to be managed, emerges as a critical relational actor in the ongoing work of commoning.

Patrick Bresnihan argues that the analytic strength of work in commoning is its capacity to "disrupt the liberal humanist epistemologies that both individualize and place humans at the centre of world-making processes" and to instead draw attention to the reality of a materially and socially constituted world of human and nonhuman interdependence, a space in which actors "navigate between limits and possibilities as they arise."[16] In this sense, sargassum and other weedy excesses become objects that challenge the boundaries of our relational world and push us toward finding "power within rather than power over" the complex ecologies of contemporary life.[17] This concern with the more than human in work on commoning articulates with similar work in anthropology that thinks of the intersecting nonhuman and human trajectories that produce the "weedy ecologies" of the Anthropocene,[18] and the flattening topography of species difference.[19] Analogous flattenings are called for in the broader multispecies work that emphasizes the emergent properties of ecological systems that, despite the strong role of humans in setting them in motion, end up decentering humans in explicit ways.[20] "Human species being," Amelia Moore points out, "is not autonomous but is instead comprised of relations with other organisms moving within geochemical processes held together by diverse forms of dependency and love."[21]

## WEEDY COLONIZATIONS OF COASTAL BELIZE

Both weeds and the sea have been objects that disrupt notions of control and a comfortable divide between nature and culture, and thus they are productive objects of analysis in anthropological scholarship and scholarship on the commons. While the sea has long been an object of meditation within anthropology, a theory machine of sorts,[22] weeds have only more recently been taken seriously.[23] As Anna Tsing points out, the Anthropocene has been defined as much by new kinds of weeds as by new kinds of technology or economic arrangements. Its landscapes are "weedy configurations," the artifacts of intersecting human and nonhuman trajectories.[24]

The seashore, on the other hand, has long been a site of interest and reflection for the social sciences and humanities, an object made enticing by its intractability (see chap. 5 for a longer discussion). The coast is a space of encounter and change defined by instability, liminality, and ontological uncertainty—a wild fluid other to the grounded culture of land. The sea has long been a site of interest, more specifically, for researchers of the commons because in that instability it has often escaped the conditions that permit easy enclosure and privatization.[25] As a result, the coastal zone has generated some of the most compelling and well-studied systems of commons management.[26] The newly weedy shore then is an ideal site for thinking about making new kinds of commons, for imagining ways of living together as commoners while still engaging the kind of difference and heterogeneity that drives and defines anthropological inquiry into the contemporary world. This notion of commoning actively engages the areas of difference in which "participant entities may come into commonality without being the same."[27] Seaweed, crabs, people, birds, small insects, and sand, among others, find themselves—human and nonhuman, animate and inanimate—living together in a new common world.

While sargassum has inundated much of the Caribbean and Gulf of Mexico, I focus ethnographically here on the small nation of Belize, which, as a small, postcolonial territory, has been frequently conceptualized historically as the difficult-to-maintain fringe of empire and capitalist penetration.[28] It has been a landscape whose people, plants, and coast have continually exceeded the political and categorical limits of colonial and capitalist projects of rationalization.[29] In a now clichéd rendering of the colony, Aldous Huxley, as an agent of British imperialism, quite unselfconsciously declared in 1934, "If the world has any ends, British Honduras would certainly be one of them."[30] He was gesturing toward the seemingly never complete, never fully knowable way British Honduras, later Belize, emerged as a colonial territory—wedged uncomfortably and ambivalently between the plantation-style economies of the British Caribbean and the well-charted Spanish territories on the rest of mainland Central America. Sargassum, which doesn't so much seek out the Belizean shoreline as get stuck there, participates unwittingly in a long colonial and postcolonial narrative of the country as an "end" that challenged the categorical distinctions that were supposed to define and regulate it.

After all, colonialism was an exercise in exerting control—between categories of people and by humans over the nonhuman. And while the commons are often imagined as an alternative to the relations of enclosure and exploitation that define colonialism and capitalism, both the commons and the colony have historically depended on similar modes of differentiation and distinction. Common-pool resource regimes are frequently romanticized by social scientists and environmentalists alike as egalitarian and eco-friendly alternatives to capitalist modes of being in the world. But scholars have increasingly called attention to the way the institutions and actors that have managed commons historically often ignored the role of social difference in constituting commons and reinforced notions of an ontological difference between people and nature.[31] Sargassum, as a weed that emerges in the watery excesses of colonial and capitalist production and exploitation, is a useful object by which the contours of new common worlds can be followed, imagined, and enacted across and through forms of social and species difference.

Huxley, who seemed to believe that no one should willingly inhabit the colonial ecology of Belize, might find some strange fascination in discovering that the same swampy ill-defined shores have become an end for the broad global circulation of a messy pelagic weed. The circulation of sargassum, which generally carries the seaweed from the east across the Lesser Antilles (or Leeward Islands) and northwesterly through the Caribbean and then eventually up through the Canal de Yucatán into the roots of the Gulf Stream, scrapes against Belize and the Yucatán Peninsula leaving behind large rafts of it that are removed from circulation and brought to shore. But, unlike in Huxley's formation, its arrival in Belize is not an end; it marks the beginning of new relations and new modes of living in common. When it washes ashore in Belize at the western end of the basin, residents and visitors can't merely find refuge on the windward side of the islands until things pass. They must confront the mess; they must get in the weeds.

## INCORPORATING THE EXCESSES OF THE ANTHROPOCENE

Sargassum is a slippery, odiferous object through which to think about the relational politics of the commons and commoning on the weedy periphery of global capitalism's Anthropocene epoch. Weeds grow on the fringes, at the spatial limits of control. They are unplanned and often undesired and thus exploit gaps within larger systems, whether in the cracks of sidewalks or, in the case of sargassum, at the newly eutrified boundaries of the ocean's shifting major current system. Weeds grow at a pace and to an extent that challenge conventional notions of temporality, territory, and relationality. In doing so they call into question conventional boundaries between nature and culture, as well as the categorical distinctions between human and nonhuman central to a geologic epoch confidently defined as dominated by a universal category of human.[32]

As middling suburban lawns frequently attest, weeds often emerge in the aftermath of poorly conceived projects of human control. Yet, more than mere symbolic objects of failure, weeds, Ashley Carse argues, can "render complex spatiotemporal relationships legible and conducive to analysis."[33] As an object that defies the conventions of capitalist control, sargassum's newly prolific existence is only made possible through the disrupted ecologies that projects of capitalist control have created (eutrophication of rivers, rising seawater temperatures, changing broad-scale ocean circulation patterns, and tourist-saturated beaches), yet in its circulation of excess, this weedy remnant reveals new relational paths in ecological and social life at a global scale. Sargassum drifts untethered to the sea floor. Its reproduction (asexual) and growth are shaped and made possible by the local environment, but it has no further allegiance to that environment; its movement is dictated by sea and, to a lesser extent, wind currents. It thus circulates through the sea without residency, responsibility, or navigational agency.

Eventually, large masses of it, sometimes miles long, unknowingly drift ashore to colonize the very Caribbean beaches that were colonized centuries earlier by European people. Ironically, this slippery new colonizer is the weedy excess of that first broad project of control and dispossession—European imperialism. Just like the colonized territories of the Caribbean, the Anthropocene ocean has been fundamentally altered in circulation and composition in the centuries that have followed.

With sargassum's arrival, people in coastal Belize and the greater Caribbean have been forced to confront how to imagine life as commoners within that weedy excess of centuries of enclosure and dispossession (see figure Con.3). Many of the people I encountered working with sargassum, often those eking out a living at the margins of a rapidly changing community, were keen to narrate the disruptive potentiality of tracing these emergent and unplanned relationships between people, environment, and economy.

Like many other liminal objects that refuse definition within binary notions of nature and culture, weeds, once unleashed, are profoundly difficult to control. Entire capitalist industries of terrestrial weed eradication have emerged in the last century. Weeds thrive in peripheral spaces, on the fringes of other more explicitly organized projects (on the sides of highways, on the edges of agricultural fields, and, most dramatically and tellingly, across vast expanses of the uncolonized seas). So perhaps the most alternative noncapitalist response to the weed would be to stop imagining its end and instead consider its incorporation.

The Anthropocene, an epochal "condition in which current ways of life (human and otherwise) are no longer able to continue their recent path," forces a reconsideration of response-ability and adaptation. Theorizing a world of persistent toxicity, Max Liboiron et al. argue that the proliferating slow disasters of climate change produce a world in which disruption is ubiquitous, but its impacts are uneven.[34] Modes of management that imagine the possibility of separation and containment, they argue, rely upon a "politics of purity that is no longer available or was never viable to begin with."[35] But in this unruly, universally disrupted commons, there is also the potential for realizing new configurations of power and new modes of living in relation to one another, to other species, and to the material environment.

FIGURE CON.3 Same Ambergris Caye beach during a sargassum inundation, 2025.

*Source*: Ben Popik

But this was not the initial inclination with the arrival of sargassum in Belize. Since the advent of the tourist industry in the country's coastal regions, tremendous labor has been invested in the cleaning of beaches to produce the immaculate, white-sand Caribbean imaginary that always eluded its beaches. Belize's shorelines benefit ecologically from the nurturing and stabilizing seagrass beds in the nearshore environment, but the rooted eelgrass also comes ashore in a regular, manageable volume that covers the desirable white sand. Each day, hoteliers and other property owners employ laborers to quietly clean the beaches in the predawn hours using a rake, pitchfork, and wheelbarrow so that tourists arrive each day to white sand. But when in 2015, largely without warning (though reports of inundations from leeward Caribbean sites were circulating), enormous rafts of sargassum began washing ashore, this job became more difficult and far more visible. The seaweed blocked access to the water, clogged boat engines, and destroyed the aesthetic ideal around which the modern commercial image of Belize (and its largest economic sector, tourism) is based. At first, nobody wanted the sargassum on their beach, yet everybody had it. Its origin could not be tracked to a single place or person, so responsibility for its arrival could never be meaningfully assessed, and responsibility for its removal emerged as a collective challenge for which there were no established collective institutions.

Institutions are at the heart of conventional theorizations of the commons. They are imagined to serve as a mediator and buffer between people and the commonly held resources. These models of commons often infer an inherent conservatism with institutions that develop and endure through well-established norms and clearly defined actors. Such models, however, risk missing much of our common world, especially the shifting and ephemeral commons that emerge through crisis, insecurity, and instability.[36] These sites of emergent commoning may prove most vital to Anthropocene life and are thus in need of ethnographic attention. It is not a coincidence that the emergent modes of commoning that occur in the face of extraordinary, and often new, types of crises encounter states and communities with weakened traditional institutions for managing them. The disastrous ecological excesses of exploitative political economies go hand in hand with the relative dearth of community and state institutions to act as a bulwark. And all this points to the vulnerabilities of resilience-oriented approaches to climate adaptation. These models often imagine institutions as modes of buffering the impacts of change rather than creative social incorporation and reorganization, the type of action possible through commoning.

Following my conversation with Roberto, he pointed out other men who had taken on the task of cleaning up the local beaches. As I began to talk and work with these men (and it is almost exclusively men performing this work), I came to see that they understood sargassum in a manner that differed markedly from those who employed them to remove it. For the property owners and town government, this was an exercise in control, an exercise in the maintenance of categories, and an exercise in the protection of capitalist value and its production in the touristic coastal zone. It was an exercise predicated on a faith in control and a relationship with nature in which agency was disproportionately a human possession. The laborers, on the other hand, often in a position of structural marginality and liminality, had no such fantasies of agentic control. In fact, they often articulated fantasies about undoing control over the space, frustrated that the beach was an increasingly policed space, particularly for minoritized others on the island. As a result, I was struck by the way the laborers began to think with and through sargassum in ways that challenged managerial notions of environmental control and narrowly capitalist notions of purity and value and recognized the vitality and relationality of this seaweed that brought its world with it as it encountered life on shore.[37]

## THE DISTRIBUTIVE ECOLOGY OF SARGASSUM

One morning, during the season of the first widespread inundation, I began talking to two men in their thirties shoveling an enormous pile of sargassum into an old, rusted Ford pickup. The pile had been slowly built, wheelbarrow load by wheelbarrow load, from the beach in front of a large expatriate home. There was a great deal of concern about the seaweed among town leaders, tourism officials, and local newspapers at that time. In that first meeting, I asked one of the men, Cairo, if he thought the sargassum was bad for San Pedro. His friend and work partner, Fernando, intervened: "Listen—I don't care—seaweed is work, man. That's all it is—work."

"And it's land, brother," Cairo added.

The seaweed they were working with had been removed from the beach a few weeks earlier, and now that it was somewhat drier, Fernando and Cairo were going to haul it to San Mateo to be used as fill. As they stood there, taking a break to talk to me, leaning against their truck, Cairo looked at me and continued, "You know people all hate the seaweed, think it's the worst thing, but for poor people like us," gesturing to Fernando, whose raised brow suggested his displeasure at the label, "it's a way to make a living."

Cairo, Fernando, and other workers I spoke with took a certain pleasure in seeing the way sargassum had made their bodies and their labor newly visible and necessary. Hotel owners, expatriates, and tourists now could see the way these workers sustained the touristic imaginary of a pristine, white sand beach and how their embodied labor created each morning the landscape they dwelled in together. More recently, the Belize Tourism Industry Association has been paying unskilled laborers $75 per day plus meals to do that work (which is a full 50 percent higher than the standard day labor rate). Seaweed then, rather than a nuisance or threat, became an object around which redistributive claims on the human economy seemed to be made by nature. Workers tasked with cleaning sargassum described how the arrival of this seaweed changed, in an admittedly limited way, the configuration of power between people, partly by highlighting the limits of human ecological control and the central role of marginalized laborers in actively producing the very places in which they were treated as peripheral. These workers remain in positions of structural inequality and vulnerability, but this weed, like a living, floating critique of the relational ecology of capitalism, created the possibility for reconfiguring relations between people and locating new spaces and forms of agency.

## CONCLUSION: DECOLONIZING THE WEEDY ANTHROPOCENE COMMONS

Efforts to respond and adapt to ecological change under climate change are often framed through the idea of resilience. Such a framing imagines that workers such as Roberto, Cairo, Fernando, and others are exhibiting a mode of resilience, an approach to precarity and instability that can be identified and replicated, one that leaves unsettled the fundamental relationships of power that produce social and ecological precarity.[38] I'm arguing here that their work represents something quite different, a mode of commoning, an actively counterhegemonic effort that takes advantage of the strange, novel ecologies of this moment to reconfigure the political economic world that they live in and to thus reconfigure the relationships of power that have come to organize the increasingly stratified social life of this growing coastal center of tourism. This is more than mere semantic distinction. It comes down fundamentally to whether these weedy new ecologies prompt us to merely engage in a mode of improvisational resilience that leaves unquestioned the conditions of precarity that necessitate such a dance or create a moment in which new relationships between people, living things, and the shifting topology of the coast make possible new ways of creating life and possibility together. These broken uncommons provide a way of expanding the imagination in a period often defined by infinite loss and pending disappearance, a period in which the urgency of that loss is often used as a foil to the imagination and possibility that might come with "an untimely questioning of the present."[39]

This is not to diminish the significance of environmental change and its urgent impacts for coastal communities in Belize. But by noticing these generative moments of common action across boundaries of bodies and species, and how these moments create zones of possibility in the messy new relational world of the Anthropocene, their work highlights the "experimental and relational character of the commons, as ongoing and variable," and the urgent need for anthropologists to observe and promote the generative work of possibility rather than retreating to the colonialist impulse to document disappearance in the face of forces such as capitalism and colonialism that we know well.[40]

Despite being relatively new to San Pedro, and the coast, Roberto employed a common narrative on the island: People there are engaged in a relationship of reciprocity with the sea, and they form a life in common with other people, organisms, and the sea itself. It was a commons held together by affective ties that linked these actors together in ways that were sustained, but also fluid, changeable, and reciprocal. And just as Marcel Mauss argued, the relationships of reciprocity yield gifts that sometimes come as an uncomfortable, or even undesirable, surprise. He argued that the importance of the gift relationship, however, was not necessarily the object exchanged but rather the relationships that the object engendered and sustained. Gifts, Mauss argued, were often rendered aggressively and represented at times great challenges to who you were, how you could relate with others, and what it might be possible for you to become.[41]

Roberto, and the other primarily poor residents of San Mateo, encountered the influx of sargassum from a position quite different than the hotel operators, town officials, tourists, or members of the media that typically approached it with a deep disgust and sense of emergency. The broad-scale forces of inequality that over time rendered a planet in common crisis—European and American imperialism, colonial dispossession, and neoliberal capitalism—placed people such as him, and regions like the Caribbean, on the fringes of emergent common understandings of the Anthropocene. In such a position, they are simultaneously the most vulnerable to this period's instability but also in a position most likely to seize on its strange productive overflows, such as the uncontainable sargassum. Their working and thinking with it to solidify land and belonging in San Pedro points to the way the Anthropocene geography is one of both enclosure and commoning, crisis and creativity. Their work points further to the need to address the differentiated reality of the Anthropocene, and the strange new commons that it may render, to rethink what it means to live—human and nonhuman—as commoners in a climactic period defined not just by disappearance and loss but by arrival and becoming and the making of new forms of value in common.

# NOTES

## INTRODUCTION: THE ENDS OF NATURE AND VALUE

1. See Emily Cooper, Lauretta Burke, and Nadia Bood, "Coastal Capital: Belize. The Economic Contribution of Belize's Coral Reefs and Mangroves," WRI working paper (World Resources Institute, 2009), http://pdf.wri.org/coastal_capital_belize_brochure.pdf.
2. See Hrishikesh Patel et al., "MIDAS: A Spatial Decision Support System for Monitoring Marine Management Areas," *International Regional Science Review* 34, no. 2 (2011): 191–214.
3. See Philip A. Kramer and Patricia R. Kramer, "Ecoregional Conservation Planning for the Mesoamerican Caribbean Reef" (World Wildlife Fund, 2002).
4. See Hannah K. Morrissette et al., "Belize Blue Carbon: Establishing a National Carbon Stock Estimate for Mangrove Ecosystems," *Science of the Total Environment* 870 (2023): 161829.
5. Assad Shoman, *Belize's Independence and Decolonization in Latin America: Guatemala, Britain, and the UN* (Palgrave Macmillan, 2010), 6.
6. Rajeshwari Dutt, *Empire on Edge: The British Struggle for Order in Belize During Yucatan's Caste War, 1847–1901* (Cambridge University Press, 2020).
7. Rico Kongsager and Esteve Corbera, "Linking Mitigation and Adaptation in Carbon Forestry Projects: Evidence from Belize," *World Development* 76 (2015): 132–46.

8. Melissa A. Johnson, "The Making of Race and Place in Nineteenth-Century British Honduras," *Environmental History* 8, no. 4 (2003): 598–617; Melissa A. Johnson, *Becoming Creole: Nature and Race in Belize* (Rutgers University Press, 2018); Laurie Kroshus Medina, "Defining Difference, Forging Unity: The Co-construction of Race, Ethnicity and Nation in Belize," *Ethnic and Racial Studies* 20, no. 4 (1997): 757–80.
9. Patrick Bresnihan, "The More-than-Human Commons: From Commons to Commoning," in *Space, Power and the Commons: The Struggle for Alternative Futures* (Routledge, 2016), 129.
10. Jason W. Moore, "The Capitalocene, Part I: On the Nature and Origins of Our Ecological Crisis," *Journal of Peasant Studies* 44, no. 3 (2017): 594–630; Heather Davis and Zoe Todd, "On the Importance of a Date, or, Decolonizing the Anthropocene," *ACME: An International Journal for Critical Geographies* 16, no. 4 (2017): 761–80; Marisol de la Cadena, Robert J. Foster, and Daniel R. Reichman, *Earth Beings: Ecologies of Practice Across Andean Worlds*, illus. ed. (Duke University Press, 2015); Elizabeth A. Povinelli, *Between Gaia and Ground: Four Axioms of Existence and the Ancestral Catastrophe of Late Liberalism* (Duke University Press, 2021).
11. Andrew S. Mathews, "Anthropology and the Anthropocene: Criticisms, Experiments, and Collaborations," *Annual Review of Anthropology* 49 (2020): 77.
12. Amelia Moore, "Anthropocene Anthropology: Reconceptualizing Contemporary Global Change," *Journal of the Royal Anthropological Institute* 22, no. 1 (2016): 33.
13. Nathan F. Sayre, "The Politics of the Anthropogenic*," *Annual Review of Anthropology* 41 (2012): 63.
14. Deborah Bird Rose, *Shimmer: Flying Fox Exuberance in Worlds of Peril* (Edinburgh University Press, 2022), 6–7.
15. Robert Costanza et al., "The Value of the World's Ecosystem Services and Natural Capital," *Nature* 387, no. 6630 (1997): 254.
16. Gretchen C. Daily, *Nature's Services* (Island, 1997).
17. Costanza et al., "The Value of the World's Ecosystem Services and Natural Capital," 254.
18. Costanza et al., "The Value of the World's Ecosystem Services and Natural Capital," 254.

19. Ernst Friedrich Schumacher, *Small Is Beautiful: Economics as If People Mattered* (Harper & Row, 1973), 15; Herman E. Daly, John B. Cobb, and Clifford W. Cobb, *For the Common Good* (Beacon, 1994), 72 Akerman, "What Does 'Natural Capital' Do? The Role of Metaphor in Economic Understanding of the Environment," *Environmental Values* 12, no. 4 (2003): 41.
20. Richard B. Norgaard, "Ecosystem Services: From Eye-Opening Metaphor to Complexity Blinder," *Ecological Economics* 69, no. 6 (2010): 1219.
21. Norgaard, "Ecosystem Services"; Vijay Kolinjivadi et al., "Neoliberal Performatives and the 'Making' of Payments for Ecosystem Services (PES)," *Progress in Human Geography* 43, no. 1 (2019): 3–25; Eli P. Fenichel and Joshua K. Abbott, "Natural Capital: From Metaphor to Measurement," *Journal of the Association of Environmental and Resource Economists* 1, no. 1/2 (2014): 1–27; Erik Gómez-Baggethun et al., "The History of Ecosystem Services in Economic Theory and Practice: From Early Notions to Markets and Payment Schemes," *Ecological Economics* 69, no. 6 (2010): 1209–18.
22. Gretchen C. Daily et al., "The Value of Nature and the Nature of Value," *Science* 289, no. 5478 (2000): 395.
23. Akerman, "What Does 'Natural Capital' Do?," 46.
24. Norgaard, "Ecosystem Services," 1226.
25. See "Moving 'Natural Capital' from Metaphor to Reality," Yale School of the Environment, May 27, 2014, http://environment.yale.edu/news/article/fenichel-moving-natural-capital-from-metaphor-to-reality/; Douglas J. McCauley, "Selling Out on Nature," *Nature* 443, no. 7107 (2006): 27–28.
26. Robert Fletcher, "Neoliberal Environmentality: Toward a Poststructuralist Political Ecology of the Conservation Debate," *Conservation and Society* 8, no. 3 (2010): 171; Suzana Sawyer, *Crude Chronicles: Indigenous Politics, Multinational Oil, and Neoliberalism in Ecuador* (Duke University Press, 2004); Sian Sullivan, "Green Capitalism, and the Cultural Poverty of Constructing Nature as Service Provider," *Radical Anthropology* 3: 18–27.
27. Patrick Bond, *Looting Africa: The Economics of Exploitation* (Zed, 2006); Karen Bakker, "Neoliberal Nature, Ecological Fixes, and the Pitfalls of Comparative Research," *Environment and Planning A* 41, no. 8 (2009): 1781–87; Bram Büscher, "Anti-Politics as Political

Strategy: Neoliberalism and Transfrontier Conservation in Southern Africa," *Development and Change* 41, no. 1 (2010): 29–51; Noel Castree, "Neoliberalising Nature: The Logics of Deregulation and Reregulation," *Environment and Planning A* 40, no. 1 (2008): 131–52; Pamela McElwee et al., "Payments for Environmental Services and Contested Neoliberalisation in Developing Countries: A Case Study from Vietnam," *Journal of Rural Studies* 36, suppl. C (2014): 423–40; Paige West, *Conservation Is Our Government Now: The Politics of Ecology in Papua New Guinea* (Duke University Press, 2006), 200; Fletcher, "Neoliberal Environmentality"; Tania Murray Li, *Land's End: Capitalist Relations on an Indigenous Frontier* (Duke University Press, 2014).

28. Stephen J. Collier, *Post-Soviet Social: Neoliberalism, Social Modernity, Biopolitics* (Princeton University Press, 2011); James Ferguson, "The Uses of Neoliberalism," *Antipode* 41, suppl. 1 (2010): 166–84; James Ferguson, *Give a Man a Fish: Reflections on the New Politics of Distribution* (Duke University Press, 2015); Ananya Roy, *Poverty Capital: Microfinance and the Making of Development* (Routledge, 2010).

29. Robert Fletcher and Jan Breitling, "Market Mechanism or Subsidy in Disguise? Governing Payment for Environmental Services in Costa Rica," *Geoforum* 43, no. 3 (2012): 402–11; Pamela D. McElwee, "Payments for Environmental Services as Neoliberal Market-Based Forest Conservation in Vietnam: Panacea or Problem?," *Geoforum* 43, no. 3 (2012): 412–26; Elizabeth Shapiro-Garza, "Contesting the Market-Based Nature of Mexico's National Payments for Ecosystem Services Programs: Four Sites of Articulation and Hybridization," *Geoforum* 46 (2013): 5–15; Maron Greenleaf, "Distributive Politics of REDD in Acre, Brazil," *Anthropology News* 55, no. 6 (2014); Maron E. Greenleaf, *Forest Lost: Producing Green Capitalism in the Brazilian Amazon* (Duke University Press, 2024).

30. Stephen J. Collier, "The Spatial Forms and Social Norms of 'Actually Existing Neoliberalism': Toward a Substantive Analytics," working paper in *International Affairs*(The New School, Graduate Program in International Affairs, 2005).

31. Patrick Gallagher and Danielle DiNovelli-Lang, "Nature and Knowledge," *Environment and Society* 5, no. 1 (2014): 1–6.

32. Tim Ingold, "Globes and Spheres: The Topology of Environmentalism," in *Environmental Anthropology: A Historical Reader*, ed. Michael Dove and Carol Carpenter (Blackwell, 2007); Fred Turner, *From Counterculture to Cyberculture: Stewart Brand, the Whole Earth Network, and the Rise of Digital Utopianism* (University of Chicago Press, 2008).
33. Joseph Masco, *The Theater of Operations: National Security Affect from the Cold War to the War on Terror* (Duke University Press, 2014).
34. Jane Poynter, *The Human Experiment: Two Years and Twenty Minutes Inside Biosphere 2* (Basic Books, 2006); Rebecca Reider, *Dreaming the Biosphere: The Theater of All Possibilities* (University of New Mexico Press, 2009).
35. Joan A. Kleypas and Chris Langdon, "Coral Reefs and Changing Seawater Carbonate Chemistry," in *Coral Reefs and Climate Change: Science and Management*, ed. Jonathan T. Phinney et al. (American Geophysical Union, 2006), 73–110.
36. Paul Ehrlich, *The Dominant Animal: Human Evolution and the Environment* (Island, 2008), 192.
37. Andrew S. Mathews, *Instituting Nature: Authority, Expertise, and Power in Mexican Forests* (MIT Press, 2011); Jamie Peck, "Liberating the City: Between New York and New Orleans," *Urban Geography* 27, no. 8 (2006): 681–713.
38. Shapiro-Garza, "Contesting the Market-Based Nature of Mexico's National Payments for Ecosystem Services Programs."
39. Robert Fletcher, "How I Learned to Stop Worrying and Love the Market: Virtualism, Disavowal, and Public Secrecy in Neoliberal Environmental Conservation," *Environment and Planning D: Society and Space* 31, no. 5 (2013): 796; Greenleaf, "Distributive Politics of REDD in Acre, Brazil."
40. James Clifford and George E. Marcus, *Writing Culture: The Poetics and Politics of Ethnography* (University of California Press, 1986), 5.
41. Clifford Geertz, "Deep Play: Notes on the Balinese Cockfight," *Daedalus* 101, no. 1 (1972): 1–37.
42. Cooper, Burke, and Bood, "Coastal Capital: Belize."
43. Johnson, *Becoming Creole*; Medina, "Defining Difference, Forging Unity."

44. Savannah Shange, *Progressive Dystopia: Abolition, Antiblackness, and Schooling in San Francisco*, illus. ed. (Duke University Press, 2019), 12.

## 1. THE ONTOLOGY OF BLUE

1. Malcom Ferdinand, Decolonial *Ecology: Thinking from the Caribbean World*, trans. Anthony Paul Smith (Polity, 2022).
2. The Welsh magistrate Basil Jones established the first cocals in Ambergris Caye in the 1790s.
3. Alejandro Camargo has written about the complex materiality of water and the resultant color-specific fluvial geography that emerges in the context of the La Mojana inland delta in Colombia. He says, "The translucence of water, therefore, dictates what can be seen or unseen and shapes people's perceptions of color. The color of water, in turn, intertwines with social notions of life and death." See Alejandro Camargo, "On the Translucence of Water," *Engagement (Anthropology and Environment)* (blog), February 8, 2024, https://aesengagement.wordpress.com/2024/02/07/on-the-translucence-of-water/.
4. It's important to note the untenability of the tourist/anthropologist distinction and the way my own imagination of that site in that moment, and my initial disappointment at being interrupted by a tourist's inquiry, was as much an instantiation of Ferdinand's critique that "the colonial perspective persists today in the tourist representation of the Caribbean as a place where one can take a break on the beach without people and offside to the world" (Ferdinand, *Decolonial Ecology*, 2).
5. Peter W. Schuhmann, "Valuation of Ecosystem Services as a Basis for Investment in Blue Economies," in *The Caribbean Blue Economy* (Routledge, 2023), 78.
6. Civil Aviation was formerly the awkward appendage to the Ministry of Tourism, which following this reorganization became the Ministry of Tourism and Civil Aviation.

   Belize is not alone in creating such a ministry. Similar government ministries have been created in Barbados, Kenya, Mauritius, and Nigeria (all former British colonies).
7. Bruno Latour, *An Inquiry Into Modes of Existence: An Anthropology of the Moderns*, trans. Catherine Porter (Harvard University Press, 2013); Zoe Todd, "An Indigenous Feminist's Take on the Ontological Turn:

'Ontology' Is Just Another Word for Colonialism," *Journal of Historical Sociology* 29, no. 1 (2016): 4–22; Eduardo Viveiros de Castro, "Cosmological Deixis and Amerindian Perspectivism," *Journal of the Royal Anthropological Institute* 4, no. 3 (1998): 469–88; David Graeber, "Radical Alterity Is Just Another Way of Saying 'Reality': A Reply to Eduardo Viveiros de Castro," *HAU: Journal of Ethnographic Theory* 5, no. 2 (2015): 1–41.

8. Todd, "An Indigenous Feminist's Take on the Ontological Turn."
9. Lucas Bessire and David Bond, "Ontological Anthropology and the Deferral of Critique," *American Ethnologist* 41, no. 3 (2014): 441.
10. Bessire and Bond, "Ontological Anthropology and the Deferral of Critique."
11. Martin Holbraad, Morten Axel Pedersen, and Eduardo Viveiros de Castro, "The Politics of Ontology: Anthropological Positions," Society for Cultural Anthropology, January 13, 2014, https://culanth.org/fieldsights/the-politics-of-ontology-anthropological-positions.
12. Bruno Latour, *We Have Never Been Modern*, trans. Catherine Porter (Harvard University Press, 1993).
13. Christopher Watts, "Relational Archaeologies: Roots and Routes," in *Relational Archaeologies* (Routledge, 2013), 1.
14. John Jervis, "The Nature of It All (Modernist Ontology)," in *Modernity Theory: Modern Experience, Modernist Consciousness, Reflexive Thinking*, ed. John Jervis (Palgrave Macmillan, 2018), 117–30.
15. Claude Lévi-Strauss, *Structural Anthropology*. (Basic Books, 1963); Lévi-Strauss was well-attuned to many of the tensions produced through a structure of binary oppositions, as has been pointed out in Carol P. MacCormack, "Nature, Culture and Gender: A Critique," in *Nature, Culture and Gender*, ed. Carol P. MacCormack and Marilyn Strathern (Cambridge University Press, 1980). Lévi-Strauss, in *The Elementary Structures of Kinship* (Beacon, 1969), made clear that the nature-culture binary is itself a cultural construction and that for him it was a methodological device. It was as much his culture as "theirs." His own mentee, Edmund Leach, went somewhat further in suggesting that "Structuralism is neither a theory or a method, but a way of looking at things." Edmund Leach, "Structuralism in Social Anthropology," in *Structuralism: An Introduction*, ed. David Robey (Clarendon, 1973), 37.

16. John Hartigan, *Aesop's Anthropology: A Multispecies Approach* (University of Minnesota Press, 2014), 4.
17. Deborah Bird Rose, *Shimmer: Flying Fox Exuberance in Worlds of Peril* (Edinburgh University Press, 2022).
18. Timothy Mitchell, *Rule of Experts: Egypt, Techno-Politics, Modernity* (University of California Press, 2002), 29.
19. Mario Blaser, "Political Ontology: Cultural Studies Without 'Cultures'?," *Cultural Studies* 23, no. 5–6 (2009): 873–96; Walker DePuy et al., "Environmental Governance: Broadening Ontological Spaces for a More Livable World," *Environment and Planning E: Nature and Space* 5, no. 2 (2022): 947–75.
20. DePuy et al., "Environmental Governance"; Sian Sullivan, "What's Ontology Got to Do with It? On Nature and Knowledge in a Political Ecology of the 'Green Economy,'" *Journal of Political Ecology* 24, no. 1 (2017): 217–42.
21. Stefan Helmreich, "Nature/Culture/Seawater," *American Anthropologist* 113, no. 1 (2011): 132.
22. Bruno Latour, *Politics of Nature: How to Bring the Sciences Into Democracy* (Harvard University Press, 2004).
23. Jacques Derrida, *Writing and Difference* (University of Chicago Press, 1978); Latour, *We Have Never Been Modern.*
24. George W. Stocking, "Franz Boas and the Culture Concept in Historical Perspective," *American Anthropologist* 68, no. 4 (1966): 867–82.
25. George W. Stocking, "From Physics to Ethnology: Franz Boas' Arctic Expedition as a Problem in the Historiography of the Behavioral Sciences," *Journal of the History of the Behavioral Sciences* 1, no. 1 (1965), 57.
26. Stocking, "From Physics to Ethnology," 59, 61.
27. Luisa Cortesi, "An Ontology of Water and Land in North Bihar, India★," *Journal of the Royal Anthropological Institute* 27, no. 4 (2021): 871.
28. Helmreich, "Nature/Culture/Seawater," 134.
29. Marshall Sahlins, *Culture and Practical Reason* (University of Chicago Press, 2013), 65.
30. Sahlins, *Culture and Practical Reason.*
31. Rebecca Solnit, *A Field Guide to Getting Lost*, reprint ed. (Penguin, 2006), 25.
32. Kenneth Little, "Blue," *Imaginativeethnography.Org* (blog), 2014, http://imaginativeethnography.org/imaginings/affect/blue/.

33. Michael Taussig, *What Color Is the Sacred?* (University of Chicago Press, 2009), 159.
34. Erin Hammeke, "Logwood Dye on Paper," 2004, 4, https://sentra.ischool.utexas.edu/~cochinea/pdfs/e-hammeke-04-logwood.pdf.
35. Eugene Newton Anderson, *Political Ecology in a Yucatec Maya Community* (University of Arizona Press, 2005), 38.
36. Hammeke, "Logwood Dye on Paper," 2–3.
37. Melissa A. Johnson, "Racing Nature and Naturalizing Race: Rethinking the Nature of Creole and Garifuna Identities," *Belizean Studies* 27, no. 2 (2005): 46.
38. O. Nigel Bolland, *Colonialism and Resistance in Belize: Essays in Historical Sociology* (Cubola Productions, 2003).
39. O. Nigel Bolland, *The Formation of a Colonial Society: Belize, from Conquest to Crown Colony* (Johns Hopkins University Press, 1977).
40. Bolland, *Colonialism and Resistance in Belize*, 31.
41. Tim Ingold, "Globes and Spheres: The Topology of Environmentalism," in *Environmental Anthropology: A Historical Reader*, ed. Michael Dove and Carol Carpenter (Blackwell, 2007).
42. Timothy Mitchell, "The World as Exhibition," *Comparative Studies in Society and History* 31, no. 2 (1989): 217–36.
43. Bolland, *The Formation of a Colonial Society*, 2.
44. Bolland, *The Formation of a Colonial Society*, 25.
45. Joel Wainwright, "'The First Duties of Persons Living in a Civilized Community': The Maya, the Church, and the Colonial State in Southern Belize," *Journal of Historical Geography* 35, no. 3 (2009): 435.
46. Timothy Choy, *Ecologies of Comparison: An Ethnography of Endangerment in Hong Kong* (Duke University Press, 2011), 13.
47. Webb Keane, "Ontologies, Anthropologists, and Ethical Life," *HAU: Journal of Ethnographic Theory* 3, no. 1 (2013): 186–91.
48. Keane, "Ontologies, Anthropologists, and Ethical Life," 188.
49. Through their growing intellectual partnerships with economists, conservation scientists seem to have picked up their propensity for imaginative base scenarios. While I've long been familiar with the economic scene setting of "Imagine that . . ." preceding some scenario of perfect information, it took longer for me to adapt to the growing propensity for the conservation scene setting of "Imagine that . . ." preceding some scenario of perfect destruction or disaster.

50. Emily Cooper, Lauretta Burke, and Nadia Bood, "Coastal Capital: Belize. The Economic Contribution of Belize's Coral Reefs and Mangroves," WRI Working Paper (World Resources Institute, 2009). http://www.wri.org/publications

## 2. NEOLIBERAL ECOLOGISTS IN OUTER SPACE

1. Gregory McNamee, *Gila: The Life and Death of an American River*, updated and expanded ed. (University of New Mexico Press, 2012).
2. For more information, see A Community on Ecosystem Services at http://www.conference.ifas.ufl.edu/aces/.
3. Joseph Alcamo et al., *Ecosystems and Human Well-Being: A Framework for Assessment* (Island, 2003).
4. The original mission (it has now been modified) of the Natural Capital Project was to "align economic forces with conservation by mainstreaming natural capital into decisions." This phrasing establishes economics and conservation as distinct and conflicting fields in need of alignment and naturalizes the existence of the object of natural capital as an object around which they will align.
5. I use the term "ecosystem service scientist" here and throughout the chapter to describe the relatively diverse set of scientific actors whose work contributes to the development of standardized tools and techniques for abstraction of ecosystem functions as "goods and services" and who frequently work to develop the means to produce monetary valuations of those goods and services. Most of these scientists are trained in specific fields such as conservation biology, ecology, and resource economics, among others. The term is not meant to suggest that those involved in the development of an ecosystem service science are a homogenous group, or even that all such scientists would even identify themselves as ecosystem service scientists. In fact, most maintain allegiance to the discipline of their training, but I use the term to emphasize the collaborative scientific community they have joined around the ideas of natural capital and ecosystem services.
6. Chunglin Kwa, "Radiation Ecology, Systems Ecology and the Management of the Environment," in *Science and Nature: Essays in the History of the Environmental Sciences*, vol. 8 (British Society for the History of Science, 1993).

7. Hugh Gusterson, *Nuclear Rites: A Weapons Laboratory at the End of the Cold War* (University of California Press, 1998); Hugh Gusterson, *People of the Bomb: Portraits of America's Nuclear Complex* (University of Minnesota Press, 2004); Joseph Masco, "'SURVIVAL IS YOUR BUSINESS': Engineering Ruins and Affect in Nuclear America," *Cultural Anthropology* 23, no. 2 (2008): 361–98; Joseph Masco, *The Nuclear Borderlands: The Manhattan Project in Post-Cold War New Mexico* (Princeton University Press, 2013); Jake Kosek, *Understories: The Political Life of Forests in Northern New Mexico* (Duke University Press, 2006); Jake Kosek, "Ecologies of Empire: On the New Uses of the Honeybee," *Cultural Anthropology* 25, no. 4 (2010): 650–78.
8. Sheila Jasanoff, *States of Knowledge: The Co-production of Science and Social Order* (Routledge, 2004); Naomi Oreskes and John Krige, *Science and Technology in the Global Cold War* (MIT Press, 2014).
9. Paul N. Edwards, *The Closed World: Computers and the Politics of Discourse in Cold War America* (MIT Press, 1997); Jacob Darwin Hamblin, *Arming Mother Nature: The Birth of Catastrophic Environmentalism* (Oxford University Press, 2013); Stuart W. Leslie, *The Cold War and American Science: The Military-Industrial-Academic Complex at MIT and Stanford* (Columbia University Press, 1993).
10. Elizabeth Shapiro-Garza, "Contesting the Market-Based Nature of Mexico's National Payments for Ecosystem Services Programs: Four Sites of Articulation and Hybridization," *Geoforum* 46 (2013): 5–15; Robert Fletcher and Jan Breitling, "Market Mechanism or Subsidy in Disguise? Governing Payment for Environmental Services in Costa Rica," *Geoforum* 43, no. 3 (2012): 402–11.; Pamela D. McElwee, "Payments for Environmental Services as Neoliberal Market-Based Forest Conservation in Vietnam: Panacea or Problem?" *Geoforum* 43, no. 3 (2012): 412–26.
11. Peder Anker, *From Bauhaus to Ecohouse: A History of Ecological Design* (Louisiana State University Press, 2010); Stephen Bocking, *Ecologists and Environmental Politics: A History of Contemporary Ecology* (Yale University Press, 1997); Joel Bartholemew Hagen, *An Entangled Bank: The Origins of Ecosystem Ecology* (Rutgers University Press, 1992); Donald Worster, *Nature's Economy: A History of Ecological Ideas* (Cambridge University Press, 1994).
12. Stephen Bocking, "Ecosystems, Ecologists, and the Atom: Environmental Research at Oak Ridge National Laboratory," *Journal of the*

*History of Biology* 28, no. 1 (1995): 1–47; Kwa, "Radiation Ecology, Systems Ecology and the Management of the Environment."

13. Anker, *From Bauhaus to Ecohouse*; Rebecca Reider, *Dreaming the Biosphere: The Theater of All Possibilities*. University of New Mexico Press, 2009.
14. Joseph Masco, "The End of Ends," *Anthropological Quarterly* 85, no. 4 (2012): 1107–24; Hamblin, *Arming Mother Nature*.
15. Peder Anker, "The Closed World of Ecological Architecture," *Journal of Architecture* 10, no. 5 (2005): 527–52.
16. Robert Costanza et al., "The Value of the World's Ecosystem Services and Natural Capital," *Nature* 387, no. 6630 (1997): 253–60.
17. Leon C. Braat and Rudolf de Groot, "The Ecosystem Services Agenda: Bridging the Worlds of Natural Science and Economics, Conservation and Development, and Public and Private Policy," *Ecosystem Services* 1, no. 1 (2012): 4–15.
18. Paul R. Ehrlich and Anne H. Ehrlich, *Extinction: The Causes and Consequences of the Disappearance of Species* (Random House, 1981).
19. This awkward dependence on the part of ecosystem service scientists on the simulation or imagination of disaster or destruction for bringing their form of value into being persists, and I discuss its problematic implications in chapter 5.
20. The "natural" descriptor marks this as a form of capital that comes from what we think of as "nature" and perhaps more problematically establishes this form of capital as completely "natural"—as something that we should have identified long before because it's always just naturally been there producing wealth.
21. Gretchen C. Daily, *Nature's Services* (Island, 1997).
22. The critic's consensus on the movie review website Rotten Tomatoes plays with the nature-society binary with which the Biosphere 2/Bio-Dome experiments, declaring acerbically "This dreadfully unfunny Pauly Shore vehicle should remain separated from society."
23. The seemingly inexplicable actions of an educational institution can often be understood by tracing donor interests, and Columbia's reluctant engagement likely had something to do with the fact that the owner of the Biosphere 2 project was Ed Bass, of the Bass oil fortune. His net worth is estimated at approximately $2 billion.
24. Thankfully, this did not happen. The University of Arizona is now effectively managing the property and using the actual Biosphere 2 enclosure as a center for innovative research on mountain hydrology.

25. Note the *Freakonomics* blog post on this topic: http://freakonomics.com/2010/01/13/the-biosphere-bubble/. There is a peculiar public fascination with narratives around the failure of the Biosphere 2 project.
26. Jane Poynter, *The Human Experiment: Two Years and Twenty Minutes Inside Biosphere 2* (Basic Books, 2006).
27. Jean Baudrillard, *The Illusion of the End* (Stanford University Press, 1994), 85.
28. Baudrillard, *The Illusion of the End*, 87.
29. And this makes it sound too simple. The Arizona Department of Agriculture prohibited the Biosphere 2 scientists from importing dirt whole from other states. Instead, they had to study its constituent elements in situ and then attempt to reproduce that dirt using elements sourced from Arizona. The making of the Biosphere 2 dirt was one of the most remarkable elements of the bizarre project.
30. Note that even in the time that I've been pondering the material for this chapter (and this is likely a sign that the process took entirely too long), these ideas of space colonization have gone from the eccentric fringe of the Biospherean past to very mainstream, perhaps even more so than during the work of the Biosphereans, because travel to Mars seems technologically feasible.
31. For a fascinating high production value iteration of the increasingly popular "They don't need us, we need them" theme in conservation, see Conservation International's star-studded campaign "Nature Is Speaking" (https://www.conservation.org/nature-is-speaking); "Nature Subsidizes Humanity with Services Worth Trillions, Ecologist Says," *Stanford News Service*, February 6, 1997, http://news.stanford.edu/pr/97/970210daily.html.
32. Dorion Sagan and Lynn Margulis, *Biospheres from Earth to Space* (Enslow, 1989), 24.
33. Baudrillard, *The Illusion of the End*, 78.
34. Jean Baudrillard, *Simulacra and Simulation* (University of Michigan Press, 1994), 78.
35. Karl Marx, *Capital: A Critique of Political Economy* (Penguin, 1976); Kathleen McAfee and Elizabeth N. Shapiro, "Payments for Ecosystem Services in Mexico: Nature, Neoliberalism, Social Movements, and the State," *Annals of the Association of American Geographers* 100, no. 3 (2010): 579–99; Kathleen McAfee, "The Contradictory Logic of Global Ecosystem Services Markets," *Development and Change* 43, no. 1 (2012): 105–31; Jessica Dempsey and Morgan M. Robertson, "Ecosystem

Services Tensions, Impurities, and Points of Engagement within Neoliberalism," *Progress in Human Geography* 36, no. 6 (2012): 758–79.

36. A. P. Kinzig et al., "Paying for Ecosystem Services—Promise and Peril," *Science* 334, no. 6056 (2011): 603–4.
37. David Graeber, *The Democracy Project: A History, a Crisis, a Movement* (Penguin, 2013); Notably, David Graeber links the stifling of alternatives to the very state system of surveillance created during the Cold War period, rapidly expanded in the post–9-11 period, and widely and enthusiastically utilized in its civilian iterations as part of the conservation movement.
38. Helga Leitner, Jamie Peck, and Eric S. Sheppard, *Contesting Neoliberalism: Urban Frontiers* (Guilford, 2007); Jamie Peck, Nik Theodore, and Neil Brenner, "Postneoliberalism and Its Malcontents," *Antipode* 41 (2009): 94–116; Shapiro-Garza, "Contesting the Market-Based Nature of Mexico's National Payments for Ecosystem Services Programs."
39. Stephen J. Collier, "The Spatial Forms and Social Norms of 'Actually Existing Neoliberalism': Toward a Substantive Analytics," working paper in *International Affairs* (The New School, Graduate Program in International Affairs, 2005); Stephen J. Collier, *Post-Soviet Social: Neoliberalism, Social Modernity, Biopolitics* (Princeton University Press, 2011); James Ferguson, "Toward a Left Art of Government: From 'Foucauldian Critique' to Foucauldian Politics," *History of the Human Sciences* 24, no. 4 (2011): 61–68; Ferguson, *Give a Man a Fish: Reflections on the New Politics of Distribution* (Duke University Press, 2015); Jean Grugel and Pía Riggirozzi, "Post-Neoliberalism in Latin America: Rebuilding and Reclaiming the State after Crisis," *Development and Change* 43, no. 1 (2012): 1–21; Aihwa Ong, "Neoliberalism as a Mobile Technology," *Transactions of the Institute of British Geographers* 32, no. 1 (2007): 3–8.
40. Karl Polanyi, *The Great Transformation: The Political and Economic Origins of Our Time* (Beacon, 1944).
41. Jamie Peck, Nik Theodore, and Neil Brenner, "Neoliberal Urbanism: Models, Moments, Mutations," *SAIS Review of International Affairs* 29, no. 1 (2009): 49–66.
42. Collier, "The Spatial Forms and Social Norms of 'Actually Existing Neoliberalism': Toward a Substantive Analytics," 12.

43. Dipesh Chakrabarty, *Rethinking Working-Class History: Bengal, 1890–1940* (Princeton University Press, 2000); J. K. Gibson-Graham, *The End of Capitalism (As We Knew It): A Feminist Critique of Political Economy* (University of Minnesota Press, 2006).
44. Andrew S. Mathews, *Instituting Nature: Authority, Expertise, and Power in Mexican Forests* (MIT Press, 2011); Jamie Peck, "Liberating the City: Between New York and New Orleans," *Urban Geography* 27, no. 8 (2006): 681–713.
45. Shapiro-Garza, "Contesting the Market-Based Nature of Mexico's National Payments for Ecosystem Services Programs."
46. Robert Fletcher, "How I Learned to Stop Worrying and Love the Market: Virtualism, Disavowal, and Public Secrecy in Neoliberal Environmental Conservation," *Environment and Planning D: Society and Space* 31, no. 5 (2013): 796; Maron Greenleaf, "Distributive Politics of REDD in Acre, Brazil," *Anthropology News* 55, no. 6 (2014).
47. Peter Galison and Bruce William Hevly, *Big Science: The Growth of Large-Scale Research* (Stanford University Press, 1992); Rebecca S. Lowen, *Creating the Cold War University: The Transformation of Stanford* (University of California Press, 1997); Cyrus C. M. Mody, "How I Learned to Stop Worrying and Love the Bomb, the Nuclear Reactor, the Computer, Ham Radio, and Recombinant DNA," *Historical Studies in the Natural Sciences* 38, no. 3 (2008): 451–61; John L. Rudolph, *Scientists in the Classroom: The Cold War Reconstruction of American Science Education* (Palgrave Macmillan, 2002); Fred Turner, *From Counterculture to Cyberculture: Stewart Brand, the Whole Earth Network, and the Rise of Digital Utopianism*, illus. ed. (University of Chicago Press, 2008).
48. Turner, *From Counterculture to Cyberculture*, 9.
49. Turner, *From Counterculture to Cyberculture*, 8.
50. Turner, *From Counterculture to Cyberculture*, 17.
51. Frank Benjamin Golley, *A History of the Ecosystem Concept in Ecology: More Than the Sum of the Parts* (Yale University Press, 1996).
52. Golley, *A History of the Ecosystem Concept in Ecology*, 3; Hagen, *An Entangled Bank*; Hamblin, *Arming Mother Nature*; Worster, *Nature's Economy*, 1994.
53. Worster, *Nature's Economy*, 372.
54. Worster, *Nature's Economy*, 372.

55. J. R. McNeill and Corinna R. Unger, "Introduction: The Big Picture," in *Environmental Histories of the Cold War*, publications of the German Historical Institute (Cambridge University Press, 2010), 13.
56. Golley, *A History of the Ecosystem Concept in Ecology*, 2.
57. Golley, *A History of the Ecosystem Concept in Ecology*, 3.
58. Golley, *A History of the Ecosystem Concept in Ecology*, 74.
59. Kwa, "Radiation Ecology, Systems Ecology and the Management of the Environment."
60. Robert Frenay, *Pulse: The Coming Age of Systems and Machines Inspired by Living Things* (Macmillan, 2006); Norbert Wiener, *Cybernetics: Or, Control and Communications in the Animal and the Machine* (Wiley, 1948).
61. Hagen, *An Entangled Bank*.
62. Paul R. Ehrlich and Peter H. Raven, "Butterflies and Plants: A Study in Coevolution," *Evolution* 18, no. 4 (1964): 586–608.
63. Ehrlich and Raven, "Butterflies and Plants," 586.
64. It is important to note, however, that Ehrlich does not actually use the term "cybernetics" in his 1963 textbook or, as best I can tell, anywhere in his published work. That said, his work is in frequent conversation, implicitly and explicitly, with thinkers who much more openly align their work with cybernetic theory. It seems hard to separate his information-oriented systems approach to biological process and relations from the very influential work in cybernetics conducted during that period.
65. Charles S. Elton, *Animal Ecology* (University of Chicago Press, 1927); A. G. Tansley, "The Use and Abuse of Vegetational Concepts and Terms," *Ecology* 16, no. 3 (1935): 284–307; Stephen Bocking, "The Ecosystem: Research and Practice in North America," *Web Ecology* 13 (2013): 43.
66. Bocking, "Ecosystems, Ecologists, and the Atom," 10; Kwa, "Radiation Ecology, Systems Ecology and the Management of the Environment," 232.
67. Turner, *From Counterculture to Cyberculture*, 44.
68. Turner, *From Counterculture to Cyberculture*, 17.
69. Anker, *From Bauhaus to Ecohouse*, 84.
70. Anker, "The Closed World of Ecological Architecture," 540.
71. This is the definition of cabin ecology from NASA: "Systems that provide for the maintenance of life in an isolated living chamber through

reutilization of the material available, in particular, by means of a cycle wherein exhaled carbon dioxide, urine, and other waste matter are converted chemically or by photosynthesis into oxygen, water, and food" (*NASA Thesaurus*, 1988), https://ntrs.nasa.gov/citations/19910007534.

72. Anker, "The Closed World of Ecological Architecture," 543; The idea of carrying capacity when applied to humans has a complicated and controversial history. See, for instance, Gretchen C. Daily and Paul R. Ehrlich, "Population, Sustainability, and Earth's Carrying Capacity," *BioScience* 42, no. 10 (1992): 761–771, for a fairly nuanced ecological consideration of carrying capacity; Gretchen C. Daily, Anne H. Ehrlich, and Paul R. Ehrlich, "Optimum Human Population Size," *Population and Environment: A Journal of Interdisciplinary Studies* 15, no. 6 (1994), for a somewhat less nuanced approach; Donella H. Meadows et al., *Limits to Growth* (Penguin, 1972) for a far less nuanced approach; and Nathan F. Sayre, "The Genesis, History, and Limits of Carrying Capacity," *Annals of the Association of American Geographers* 98, no. 1 (2008): 120–134, for a fascinating social history of the term and its usage from the perspective of a cultural geographer. Anker's "The Closed World of Ecological Architecture" claims that the first contemporary usage of the term was by Mark Twain to discuss a ship's maximum cargo load in 1883. This would offer great narrative potential, but Sayre's detailed history suggests it isn't the case.
73. Kenneth Boulding, "The Economics of the Coming Spaceship Earth," in *Environmental Quality in a Growing Economy*, ed. Henry Jarrett (Resources for the Future Press, 1966), 3–14.
74. Anker, *From Bauhaus to Ecohouse*, 97.
75. W. B. Cassidy, ed., *Bioengineering and Cabin Ecology*, vol. 20 of the Science and Technology series (American Astronautical Society, 1969), 1.
76. Stefan Helmreich, "From Spaceship Earth to Google Ocean: Planetary Icons, Indexes, and Infrastructures," *Social Research: An International Quarterly* 78, no. 4 (2011): 1211–42.
77. Stewart Brand, "We Are as Gods," in *Whole Earth Catalog* (Portola Institute, 1968).
78. Richard Buckminster Fuller, *Operating Manual for Spaceship Earth* (Simon and Schuster, 1970), 47.
79. Fuller, *Operating Manual for Spaceship Earth*, 53.

80. Eugene Pleasants Odum and Gary W. Barrett, *Fundamentals of Ecology* (Thomson Brooks/Cole, 2005), 134.
81. "The Natural Capital Project," Stanford Woods Institute for the Environment, 2015, https://woods.stanford.edu/research/centers-programs/natural-capital-project.
82. Henry George, *Progress and Poverty: An Inquiry Into the Cause of Industrial Depressions and of Increase of Want with Increase of Wealth; the Remedy* (National Single Tax League, 1879).
83. W. Patrick McCray, *The Visioneers: How a Group of Elite Scientists Pursued Space Colonies, Nanotechnologies, and a Limitless Future* (Princeton University Press, 2012), 21.
84. Masco, "The End of Ends"; Masco, *The Nuclear Borderlands*.
85. Hannah Arendt, *Men in Dark Times* (Mariner, 1970), 83.
86. Faisal Devji, *The Terrorist in Search of Humanity: Militant Islam and Global Politics* (Columbia University Press, 2008), 11.
87. Paul Ehrlich, *The Population Bomb* (Ballantine, 1968).
88. Meadows et al., *Limits to Growth*.
89. Sara Holiday Nelson, "Beyond the Limits to Growth: Ecology and the Neoliberal Counterrevolution," *Antipode* 47, no. 2 (2015): 467.
90. Meadows et al., *Limits to Growth*.
91. The distinction between predictive models and scenario models is emphasized in ecosystem service science as well and is a distinction that emerged in Cold War–era planning for atomic disaster. Credit for the development of contemporary scenario planning is often given to Herman Kahn of the quintessential Cold War institution, the RAND Corporation. Fittingly, scenario planning, originally intended to prepare for the devastation of war, would transfer as a technology smoothly to large corporations, most notably the Shell Corporation. Many large corporate actors from the consultancy firm McKinsey to American Express now advocate for it (Liam Fahey and Robert M. Randall, *Learning from the Future* [Wiley, 1998], 10).
92. Paul N. Edwards, *A Vast Machine: Computer Models, Climate Data, and the Politics of Global Warming* (MIT Press, 2010), 369.
93. Rob Krueger and David Gibbs, *The Sustainable Development Paradox: Urban Political Economy in the United States and Europe* (Guilford, 2007), 44; Egon Becker and Thomas Jahn, "Growth or Development," in *Political Ecology: Global and Local*, ed. Roger Keil (Routledge, 1998), 68–86.

94. Richard H. Grove, *Green Imperialism: Colonial Expansion, Tropical Island Edens and the Origins of Environmentalism, 1600–1860* (Cambridge University Press, 1996); Peder Anker, *Imperial Ecology: Environmental Order in the British Empire, 1895–1945* (Harvard University Press, 2001); Timothy Mitchell, *Rule of Experts: Egypt, Techno-Politics, Modernity* (University of California Press, 2002).
95. Cameron Walker, "Ecosystem Marketplace—Conservation Central's The Daily Show," Ecosystem Marketplace, 2006.
96. Hali Healy, Sylvia Lorek, and Beatriz Rodríguez-Labajos, "Ecosystem Services," EJOLT (Environmental Justice Organisations, Liabilities and Trade), accessed July 30, 2025, http://www.ejolt.org/2013/05/ecosystem-services/.

## 3. FISH STORIES: CIRCULATING NARRATIVES OF RACE AND COLONIALISM

1. Karl Marx, *Wages and Capital* (Labour Literature Society, 1893), 11.
2. Yarimar Bonilla, "Unsettling Sovereignty," *Cultural Anthropology* 32, no. 3 (2017): 330–39, has written extensively about the way that sovereignty functioned as an imperial legal technology as opposed to the description of any natural rights to territory.
3. Morgan M. Robertson, "The Nature That Capital Can See: Science, State, and Market in the Commodification of Ecosystem Services," *Environment and Planning D: Society and Space* 24, no. 3 (2006): 367–87; Elizabeth Shapiro-Garza, "Contesting the Market-Based Nature of Mexico's National Payments for Ecosystem Services Programs: Four Sites of Articulation and Hybridization," *Geoforum* 46 (2013): 5–15; Sian Sullivan, "Banking Nature? The Spectacular Financialisation of Environmental Conservation," *Antipode* 45, no. 1 (2013): 198–217.
4. Eric Williams, *Capitalism and Slavery*, unknown ed. (University of North Carolina Press, 1994).
5. Melissa Johnson, "The Making of Race and Place in Nineteenth-Century British Honduras," *Environmental History* 8, no. 4 (2003): 598–617; Melissa Johnson, *Becoming* Creole: *Nature and Race in Belize* (Rutgers University Press, 2018).
6. Robert Costanza et al., "The Value of the World's Ecosystem Services and Natural Capital," *Nature* 387, no. 6630 (1997); Gretchen Daily,

*Nature's Services* (Island, 1997); Peter Kareiva et al., *Natural Capital: Theory and Practice of Mapping Ecosystem Services* (Oxford University Press, 2011).

7. Paige West, *Conservation Is Our Government Now: The Politics of Ecology in Papua New Guinea* (Duke University Press, 2006); Pamela D. McElwee, "Payments for Environmental Services as Neoliberal Market-Based Forest Conservation in Vietnam: Panacea or Problem?," *Geoforum* 43, no. 3 (2012); James Igoe and Dan Brockington, "Neoliberal Conservation: A Brief Introduction," *Conservation and Society* 5, no. 4 (2007): 232–49; Arun Agrawal, *Environmentality: Technologies of Government and the Making of Subjects* (Duke University Press, 2005); Paige West, *Dispossession and the Environment: Rhetoric and Inequality in Papua New Guinea* (Columbia University Press, 2016); James Ferguson, "Seeing Like an Oil Company: Space, Security, and Global Capital in Neoliberal Africa," *American Anthropologist* 107, no. 3 (2005): 377–82.
8. Laurie Kroshus Medina, "Defining Difference, Forging Unity: The Co-construction of Race, Ethnicity and Nation in Belize," *Ethnic and Racial Studies* 20, no. 4 (1997): 758.
9. Medina, "Defining Difference, Forging Unity," 758.
10. O. Nigel Bolland, *The Formation of a Colonial Society: Belize, from Conquest to Crown Colony* (Johns Hopkins University Press, 1977); Assad Shoman, *Belize's Independence and Decolonization in Latin America: Guatemala, Britain, and the UN* (Palgrave Macmillan, 2010).
11. O. Nigel Bolland, *Colonialism and Resistance in Belize: Essays in Historical Sociology* (Cubola Productions, 2003), 123.
12. Michael Garfield Smith, *The Plural Society in the British West Indies* (University of California Press, 1965), 310.
13. Bolland, *Colonialism and Resistance in Belize*, 210; Medina, "Defining Difference, Forging Unity."
14. Ann Laura Stoler, *Carnal Knowledge and Imperial Power: Race and the Intimate in Colonial Rule* (University of California Press, 2010), xiii.
15. Krista A. Thompson, *An Eye for the Tropics: Tourism, Photography, and Framing the Caribbean Picturesque* (Duke University Press, 2007) Mimi Sheller, *Consuming the Caribbean: From Arawaks to* Zombies (Routledge, 2003).

16. Emily Cooper, Lauretta Burke, and Nadia Bood, "Coastal Capital: Belize. The Economic Contribution of Belize's Coral Reefs and Mangroves," WRI Working Paper (World Resources Institute, 2009).
17. Dan D. Brockington, *Fortress Conservation: The Preservation of the Mkomazi Game Reserve, Tanzania* (Indiana University Press, 2002).
18. Karen Bakker, "Neoliberal Nature, Ecological Fixes, and the Pitfalls of Comparative Research," *Environment and Planning A* 41, no. 8 (2009): 1781–87; Noel Castree, "Neoliberalising Nature: The Logics of Deregulation and Reregulation," *Environment and Planning A* 40, no. 1 (2008): 131–52; James McCarthy and Scott Prudham, "Neoliberal Nature and the Nature of Neoliberalism," *Geoforum* 35, no. 3 (2004): 275–83; Sullivan, "Banking Nature?," 198–217; McElwee, "Payments for Environmental Services as Neoliberal Market-Based Forest Conservation in Vietnam," 412–26.
19. Costanza et al., "The Value of the World's Ecosystem Services and Natural Capital," 254.
20. Erik Gómez-Baggethun et al., "The History of Ecosystem Services in Economic Theory and Practice: From Early Notions to Markets and Payment Schemes," *Ecological Economics* 69, no. 6 (2010): 1209–18; Richard B. Norgaard, "Ecosystem Services: From Eye-Opening Metaphor to Complexity Blinder," *Ecological Economics* 69, no. 6 (2010): 1219–27; Shapiro-Garza, "Contesting the Market-Based Nature of Mexico's National Payments for Ecosystem Services Programs," 5–15.
21. Johnson, "The Making of Race and Place in Nineteenth-Century British Honduras," 598–617; Henri Lefebvre, *The Production of Space* (Wiley-Blackwell, 1991).
22. Jose Sanchez, "How Will San Pedro Get Rid of Undesirables on the Island?," March 19, 2010, https://edition.channel5belize.com/archives/29866.
23. John C. Everitt, "The Torch Is Passed: Neocolonialism in Belize," *Caribbean Quarterly* 33, no. 3/4 (1987): 42–59.
24. Michael Cutler Stone, *Caribbean Nation, Central American State: Ethnicity, Race, and National Formation in Belize, 1798–1990* (University of Texas, 1994).
25. The anxiety surrounding this category can be felt in phrases like the "Latinization of Belize," the zeitgeist of which is captured in the 2012

*Economist* article "Immigrants in Belize: A Spanish Accent," http://www.economist.com/node/21547848.

26. Angel Nuñez, "25 Years Ago in San Pedro-Garifunas in San Pedro," Ambergris Today, November 17, 2005.
27. Produce is another scarce resource on the island, but its scarcity is a bit more easily explained by the lack of local agricultural production.
28. This price compares favorably with local market prices elsewhere in the Caribbean region. However, it marks a substantial increase in relative price for Belize, and especially for San Pedro, where until the last twenty years lobster was largely available at low prices or through relationships of reciprocity with family members and friends. Fewer people now have access to these provisioning relations. The daily wage for a working-class person in San Pedro is about US $20/day, meaning that the price of a single lobster tail approaches their daily wage.
29. Roberto Pott, "Evaluating Local Access to Seafood in Urban Belize City" ((master's diss., Duke University, 2010).
30. I use Arquel's real name here because he wanted me to and, sadly, he passed away far too early in 2014. Other names in this chapter are pseudonyms.
31. Beatrice Crona et al., "Middlemen, a Critical Social-Ecological Link in Coastal Communities of Kenya and Zanzibar," *Marine Policy* 34, no. 4 (2010): 766.
32. Sara M. Lewis, "The Role of Herbivorous Fishes in the Organization of a Caribbean Reef Community," *Ecological Monographs* 56, no. 3 (1986): 183–200; Peter J. Mumby, "The Impact of Exploiting Grazers (Scaridae) on the Dynamics of Caribbean Coral Reefs," *Ecological Applications* 16, no. 2 (2006): 747–69; Peter J. Mumby, Alan Hastings, and Helen J. Edwards, "Thresholds and the Resilience of Caribbean Coral Reefs," *Nature* 450, no. 7166 (2007): 98–101.
33. Courtney E. Cox et al., "Genetic Testing Reveals Some Mislabeling but General Compliance with a Ban on Herbivorous Fish Harvesting in Belize," *Conservation Letters* 6, no. 2 (2013): 132–40.
34. However, there is a definite relationship between these activities and the actual geography of George Street. To that end, the UK Foreign Office, in its Belize advisory, specifically warns against traveling in "the George Street area."

35. Johnson, "The Making of Race and Place in Nineteenth-Century British Honduras," 598.
36. "The Natural Capital Alliance—Marine InVEST in Belize," accessed 18 March 2015, http://www.naturalcapitalproject.org/pubs/marine/MarineInVEST_Belize.pdf.
37. Johnson, "The Making of Race and Place in Nineteenth-Century British Honduras"; Johnson, *Becoming Creole.*

## 4. GROUND TRUTHING: THE HUMAN LABOR OF NATURAL CAPITAL

1. I thank Jessica Barnes for pointing out the complex and conflicted history of the term "jungle" and for pressing me to critically consider its usage here Michael R. Dove, "The Dialectical History of 'Jungle' in Pakistan: An Examination of the Relationship Between Nature and Culture," SSRN Scholarly Paper (Social Science Research Network, 1992), http://papers.ssrn.com/abstract=2403490; Francis Zimmermann, *Jungle and the Aroma of Meats: An Ecological Theme in Hindu Medicine* (Motilal Banarsidass, 2011). I use the term because it is the term most frequently used by both scientists and Indigenous peoples in the conversations that I had regarding the study space.
2. The name, rather than a critique of the village that has grown at the intersection, refers instead to the practice of dumping fill material to create a roadway or causeway through low-lying wetlands in Belize.
3. The Mennonites in Belize, roughly ten thousand of them, are primarily Russian Mennonites, the descendants of German-speaking Anabaptists who migrated to the Vistula Delta of the then–Russian Empire in the late eighteenth century. They began migrating to countries in the Western Hemisphere in the 1900s, many moving to Canada, and a portion of that population eventually resettled to Central America, including Belize, in the middle of the twentieth century. Russian Mennonites generally speak a form of Low German called Plautdietsch, but in Belize many Mennonites also speak English, and in some cases Kriol or Spanish, depending on their level of engagement with broader society. Mennonite communities in Belize range from highly conservative and traditional, living in relative isolation from the rest of Belizean

society, to highly socially integrated operators of large industrial farms and building operations. The Mennonites of Blue River, a pseudonym, are among the most conservative and traditional of the Belizean Mennonite communities.

4. To protect the identity of the workers and their collaborators, this is a pseudonym.
5. "Breed" is often used as a casual alternate verb for "impregnate", as in "He breed his girlfriend again."
6. Full disclosure: I was wearing essentially the same thing. We were both "in the field."
7. "The Global Positioning System," GPS.gov, accessed June 21, 2015, http://www.gps.gov/systems/gps/.
8. James Aronson, Suzanne J. Milton, and James N. Blignaut, *Restoring Natural Capital: Science, Business, and Practice* (Island, 2012); Peter Kareiva et al., *Natural Capital: Theory and Practice of Mapping Ecosystem Services* (Oxford University Press, 2011).
9. Helen D. Hazen and Leila M. Harris, "Limits of Territorially-Focused Conservation: A Critical Assessment Based on Cartographic and Geographic Approaches," *Environmental Conservation* 34, no. 4 (2007): 280–90; Diana M. Liverman and Rosa Maria Roman Cuesta, "Human Interactions with the Earth System: People and Pixels Revisited," *Earth Surface Processes and Landforms* 33, no. 9 (2008): 1458–71.
10. Noel Castree, "Neoliberalism and the Biophysical Environment 1: What 'Neoliberalism' Is, and What Difference Nature Makes to It," *Geography Compass* 4, no. 12 (2010): 1725–33; Robert Costanza, "Valuing Natural Capital and Ecosystem Services Toward the Goals of Efficiency, Fairness, and Sustainability," *Ecosystem Services* 43 (2020): 101096; Kareiva et al., *Natural Capital*; Robert Fletcher, *Failing Forward: The Rise and Fall of Neoliberal Conservation* (University of California Press, 2023).
11. Michel Callon, "Some Elements of a Sociology of Translation: Domestication of the Scallops and the Fishermen of Saint Brieuc Bay," in *The Science Studies Reader* (Routledge, 1999).
12. Sheila Jasanoff, "Image and Imagination: The Formation of Global Environmental Consciousness," in *Changing the Atmosphere: Expert Knowledge and Environmental Governance* (MIT Press, 2001); Myanna Lahsen, "Seductive Simulations? Uncertainty Distribution Around

Climate Models," *Social Studies of Science* 35, no. 6 (2005): 895–922; Naomi Oreskes, Kristin Shrader-Frechette, and Kenneth Belitz, "Verification, Validation, and Confirmation of Numerical Models in the Earth Sciences," *Science* 263, no. 5147 (1994): 641–46.

13. Anna Lowenhaupt Tsing, Andrew S. Mathews, and Nils Bubandt, "Patchy Anthropocene: Landscape Structure, Multispecies History, and the Retooling of Anthropology: An Introduction to Supplement 20," *Current Anthropology* 60, no. S20 (2019): S186–97; Eduardo Viveiros de Castro, "On Models and Examples: Engineers and Bricoleurs in the Anthropocene," *Current Anthropology* 60, no. S20 (2019): S296–308.
14. Naomi Oreskes, Kristin Shrader-Frechette, and Kenneth Belitz, "Verification, Validation, and Confirmation of Numerical Models in the Earth Sciences," *Science* 263, no. 5147 (1994): 642.
15. Daniel Miller, "Turning Callon the Right Way Up," *Economy and Society* 31, no. 2 (2002): 218–33.
16. Donald Mackenzie, "Is Economics Performative? Option Theory and the Construction of Derivatives Markets," *Journal of the History of Economic Thought* 28, no. 1 (2006): 29–55.
17. Micha Rahder, *An Ecology of Knowledges: Fear, Love, and Technoscience in Guatemalan Forest Conservation* (Duke University Press, 2020).
18. Lahsen, "Seductive Simulations?," 906, 911.
19. Jason Danely, "Affect, Infrastructure, and Vulnerability: Making and Breaking Japanese Eldercare," *Medicine Anthropology Theory* 3, no. 2 (2016): 198; James G. Carrier and Daniel Miller, *Virtualism: A New Political Economy* (Berg, 1998); James G. Carrier and Paige West, *Virtualism, Governance and Practice* (Berghahn, 2009).
20. Geoffrey C. Bowker, "Time, Money, and Biodiversity," in *Global Assemblages*, ed. Aihwa Ong and Stephen J. Collier (Blackwell, 2005), 107.
21. Bowker, "Time, Money, and Biodiversity," 107.
22. While there has been limited research to date on the role of affect in shaping digital conservation labor, Stuart C. Aitken and Mei-Po Kwan, "GIS as Qualitative Research: Knowledge, Participatory Politics and Cartographies of Affect," in *The SAGE Handbook of Qualitative Geography*, ed. Dydia DeLyser, Steve Herbert, Stuart Aitken, Mike Crang, and Linda McDowell (SAGE, 2010), offers a compelling review of the role of affect in GIS and cartography more broadly.

23. Sarah Besky and Alex Blanchette, eds., *How Nature Works: Rethinking Labor on a Troubled Planet* (University of New Mexico Press published in association with the School for Advanced Research Press, 2019); Ashley Carse, "Nature as Infrastructure: Making and Managing the Panama Canal Watershed," *Social Studies of Science* 42, no. 4 (2012): 539–63.
24. Brian Larkin, "The Politics and Poetics of Infrastructure," *Annual Review of Anthropology* 42, no. 1 (2013): 328.
25. Nikhil Anand, "Pressure: The PoliTechnics of Water Supply in Mumbai," *Cultural Anthropology* 26, no. 4 (2012): 542–64; Nikhil Anand, "Leaky States: Water Audits, Ignorance, and the Politics of Infrastructure," *Public Culture* 27, no. 2 (76) (2015): 305–30; Hannah Appel and Mukul Kumar, "Finance," Society for Cultural Anthropology, September 24, 2015, https://culanth.org/fieldsights/717-finance; Carse, "Nature as Infrastructure."
26. Nikhil Anand, Akhil Gupta, and Hannah Appel, eds., *The Promise of Infrastructure* (Duke University Press, 2018), 3.
27. Steven J. Jackson et al., "Understanding Infrastructure: History, Heuristics and Cyberinfrastructure Policy," *First Monday* 12, no. 6 (2007).
28. Nikhil Anand, *Hydraulic City: Water and the Infrastructures of Citizenship in Mumbai* (Duke University Press, 2017).
29. Jackson et al., "Understanding Infrastructure."
30. Donald Worster, *Nature's Economy: The Roots of Ecology* (Sierra Club, 1977).
31. Larkin, "The Politics and Poetics of Infrastructure," 328.
32. Anand, Gupta, and Appel, *The Promise of Infrastructure*; Appel and Kumar, "Finance";; Larkin, "The Politics and Poetics of Infrastructure."
33. Kathleen Stewart, *Ordinary Affects* (Duke University Press, 2007), 2–3.
34. Jerry Jacka, "Place, Time, and Affect: Changing Landscapes Around a New Guinea Mining Area," *Cultural Anthropology* 37, no. 3 (2022): 549–71; Emily Martin, "The Potentiality of Ethnography and the Limits of Affect Theory," *Current Anthropology* 54, no. S7 (2013): S149–58; Patricia Ticineto Clough and Jean Halley, eds., *The Affective Turn: Theorizing the Social* (Duke University Press, 2007); Danilyn Rutherford, "Affect Theory and the Empirical," *Annual Review of Anthropology* 45, no. 1 (2016): 285–300.

35. Laura Bear et al., "Gens: A Feminist Manifesto for the Study of Capitalism," Society for Cultural Anthropology, March 30, 2015, https://culanth.org/fieldsights/652-gensa-feminist-manifesto-for-the-study-of-capitalism.
36. Bear et al., "Gens."
37. Neera M. Singh, "The Affective Labor of Growing Forests and the Becoming of Environmental Subjects: Rethinking Environmentality in Odisha, India," *Geoforum* 47 (2013): 189–98; Neera M. Singh, "Introduction: Affective Ecologies and Conservation," *Conservation and Society* 16, no. 1 (2018): 1–7; Marien González-Hidalgo and Christos Zografos, "Emotions, Power, and Environmental Conflict: Expanding the 'Emotional Turn' in Political Ecology," *Progress in Human Geography* 44, no. 2 (2020): 235–55.
38. Of course, notions of "true color" are problematic for their own reasons, a problem well examined in anthropology. See chapter 1 for a discussion of Boas's study of the color blue and the importance of the differential perception of color in shaping early American anthropological inquiry.
39. S. C. Liew, "Optical Remote Sensing (Tutorial)," Center for Remote Image Sensing and Processing (CRISP), 2001, http://www.crisp.nus.edu.sg/~research/tutorial/optical.htm#spectral_signature.
40. The relationship between various types of land cover and their functional value for processes such as carbon sequestration is of course complicated and subject to extensive debate. See, for example, Michel Callon, "Civilizing Markets: Carbon Trading Between *in Vitro* and *in Vivo* Experiments," *Accounting, Organizations and Society* 34, no. 3–4 (2009): 535–48; Donald MacKenzie, "Making Things the Same: Gases, Emission Rights and the Politics of Carbon Markets," *Accounting, Organizations and Society* 34, no. 3–4 (2009): 440–55.
41. "Confusion Matrix—GIS Wiki | The GIS Encyclopedia," accessed January 14, 2019, http://wiki.gis.com/wiki/index.php/Confusion_matrix.
42. Stephen V. Stehman, "Selecting and Interpreting Measures of Thematic Classification Accuracy," *Remote Sensing of Environment* 62, no. 1 (1997): 77.
43. Global Land et al., "Global Land Cover Validation: Recommendations for Evaluation and Accuracy Assessment of Global Land Cover Maps" (Office for Official Publications of the European Communities, 2006), 12, http://landval.gsfc.nasa.gov/pdf/GlobalLandCoverValidation.pdf.

44. Giles M. Foody, "Status of Land Cover Classification Accuracy Assessment," *Remote Sensing of Environment* 80, no. 1 (2002): 194.
45. J. Velásquez Runk et al., "Landscapes, Legibility, and Conservation Planning: Multiple Representations of Forest Use in Panama," *Conservation Letters* 3, no. 3 (2010): 167–76, further demonstrated that Landsat satellite imagery tends to obscure key human uses of the landscape and that the "selective harvest of large trees, most of the rice swiddens, and one homegarden site were not distinguishable from mature forest using Landsat satellite imagery and broad supervised classification methods" (172). Their findings emphasize that even "correct" classification fails to read important aspects of the social landscape and can contribute to impoverished understandings of the relationship between environment and society.
46. Hugh Raffles, *In Amazonia: A Natural History* (Princeton University Press, 2002), 169.
47. Tim Ingold, "Globes and Spheres: The Topology of Environmentalism," in *Environmental Anthropology: A Historical Reader*, ed. Michael Dove and Carol Carpenter (Blackwell, 2007), 469.
48. Jean Baudrillard, *Simulacra and Simulation* (University of Michigan Press, 1994); Carrier and West, *Virtualism, Governance and Practice*; Guy Debord, *The Society of the Spectacle* (Zone, 1994); Jim Igoe, Katja Neves, and Dan Brockington, "A Spectacular Eco-Tour Around the Historic Bloc: Theorising the Convergence of Biodiversity Conservation and Capitalist Expansion," *Antipode* 42, no. 3 (2010): 486–512.
49. Karen Ho, *Liquidated: An Ethnography of Wall Street* (Duke University Press, 2009).
50. It's worth noting that their position bears resemblance to that of the colonial and development official of the past and present—sent from imagined core to imagined periphery to enact programs of development and rationalization, with which they have varying degrees of actual affinity.
51. Mestizo, as an ethnic category in Belize, most commonly refers to people of mixed Spanish and Indian heritage, most of whom migrated south into current-day Belize during the long and horrifically violent Caste War of Yucatán (*La Guerra de las Castas*) in the second half of the nineteenth century. Confusingly, and somewhat problematically,

the category has often come to describe all people who would consider themselves "Spanish," so it now includes virtually any Central American migrant as well. The history and politics of racial and ethnic identification in Belize are discussed in greater detail in chap. 2.

52. However, in some cases, speakers would strategically turn to exclusively Spanish in key moments as a political gesture. This was especially common on San Pedro as native San Pedranos would attempt to assert a Mestizoness to San Pedro culture.
53. Osmer Balam, "Overt Language Attitudes and Linguistic Identities Among Multilingual Speakers in Northern Belize," *Studies in Hispanic and Lusophone Linguistics* 6, no. 2 (2013): 249.
54. There is a literature on these forms of translation in other technical fields, such as medicine. Lucía Ruiz Rosendo, "English and Spanish Medical Languages: A Comparative Study from Translation Point of View," *TRANS: Revista de Traductología*, no. 12 (2008): 231–46.
55. Bruno Latour, in his review of Shapin and Schaffer, discusses the long history of the unmarked deskilling of scientific labor as part of the effort to make particular ways of knowing appear easily replicable and universal. Discussing the widespread adoption of the air pump as a piece of standard laboratory equipment, he says, "Instead of requiring major investment and great skill deployed by big scientists the pump may now be activated with little money and little competence by little scientists. This erosion of skills, this displacement in the point of application of the know—how, this fascinating way through which talked about instruments become silent pieces of equipment, this shift from physicists to instrument makers, from Ph.D.s to technicians, is not well captured by notions such as practice or skill, since what is to be explained is a redistribution and reorganisation of skills. The skills that were necessary at the beginning of the century are no longer necessary at the end. They have been delegated to reskilled non-humans." Latour, "Postmodern? No, Simply Amodern! Steps towards an Anthropology of Science," *Studies in History and Philosophy of Science* Part A 21, no. 1 (1990): 145–71.
56. I discuss the history and work of the Natural Capital Project and its lead scientists in more detail in chap. 2 as part of a discussion of the intellectual history of the "natural capital" and "ecosystem services" concepts.

57. David DiBiase, "Strengthening the GIS Profession," *ArcNews*, 2012, http://www.esri.com/news/arcnews/summer12articles/strengthening-the-gis-profession.html.
58. Jung Eun Hong, "Identifying Skill Requirements for GIS Positions: A Content Analysis of Job Advertisements," *Journal of Geography* 115, no. 4 (2016): 147–58; Harlan Onsrud and Werner Kuhn, *Advancing Geographic Information Science: The Past and Next Twenty Years* (Lulu.com, 2016); B. A. Ricker et al., "Tool, Toolmaker, and Scientist: Case Study Experiences Using GIS in Interdisciplinary Research," *Cartography and Geographic Information Science* 47, no. 4 (2020): 350–66.
59. Nancy J. Obermeyer, "GIS: The Maturation of a Profession," *Cartography and Geographic Information Science* 34, no. 2 (2007): 129–32.
60. Caitlyn Dempsey, "The GIS Career, Revisioned," *GIS Lounge* (blog), 2012, http://www.gislounge.com/the-gis-career-revisioned/.
61. Caitlin Dempsey, "GIS Job Definitions," *GIS Lounge* (blog), April 17, 2022, https://www.gislounge.com/gis-job-definitions/.
62. There has been a proliferation of open source and web-based programs for geospatial software and data over the last decade, but many large organizations and government agencies have been slow to move away from established for-profit providers despite their very high licensing fees and sometimes restrictive programs. See Coetzee et al., "Open Geospatial Software and Data: A Review of the Current State and A Perspective into the Future," *ISPRS International Journal of Geo-Information* 9, no. 2 (2020): 90, for a review of the possibilities and challenges of open geospatial data and software.
63. Raffles, *In Amazonia*, 171.
64. Raffles, *In Amazonia*, 171.
65. Ingold, "Globes and Spheres."
66. Ingold, "Globes and Spheres," 465.

## 5. THE POLITICAL ECOLOGY OF THE SEAWALL

1. This payment far exceeded the minimum hourly wage in Belize at the time, about BZ$3, and was even on the high side for Ambergris Caye, where wages, and the cost of living, are generally higher. But it was still

a brutal, wet, salty form of labor for a low wage. Wages, back to the coconut plantations, have generally exceeded Belize minimum wages, but working on Ambergris Caye then and now has come at great cost to laborers because of the high cost of housing and food and in many cases separation from their mainland family.

2. I use the term "migrant" here intentionally rather than the term "expatriate," which is commonly used to describe the primarily white North American and European migrants to Belize. "Expatriate" is a term that solidifies the privileged modes of circulation enjoyed by some migrants and marks their movement as unproblematic and desirable in contrast to typical migrants. Adia Benton, "African Expatriates and Race in the Anthropology of Humanitarianism," *Critical African Studies* 8, no. 3 (2016): 266–77; Erik Cohen, "Expatriate Communities," *Current Sociology* 24, no. 3 (1977): 5–90. In other places in the text, I use "expatriate" to highlight the terms under which these migrants understand themselves or how others identify them.
3. Mixed and misplaced colonial symbols run rampant in places like San Pedro, products of a specific colonial history that endures through a more generalized fetishization of colonial symbols and architectural forms. "Tiki"-style cabanas abut southern US–style plantation cabins, Mesoamerican *palapas* sit at the end of powerboat docks, and high-end resorts reproduce imagined British colonial architectural motifs (see introduction).
4. Kathryn Yusoff, "Geologic Realism: On the Beach of Geologic Time," *Social Text* 37, no. 1 (138) (2019): 12.
5. Édouard Glissant, *Poetics of Relation*, trans. Betsy Wing (University of Michigan Press, 1997).
6. Cara Daggett, "Petro-Masculinity: Fossil Fuels and Authoritarian Desire," *Millennium* 47, no. 1 (2018): 42.
7. See, for example, Nikhil Anand, *Hydraulic City: Water and the Infrastructures of Citizenship in Mumbai* (Duke University Press, 2017); Nikhil Anand, Akhil Gupta, and Hannah Appel, eds., *The Promise of Infrastructure* (Duke University Press, 2018); Sarah E. Vaughn, *Engineering Vulnerability: In Pursuit of Climate Adaptation* (Duke University Press, 2022); Brian Larkin, "The Politics and Poetics of Infrastructure," *Annual Review of Anthropology* 42, no. 1 (2013): 327–43; Hannah Appel,

*The Licit Life of Capitalism: U.S. Oil in Equatorial Guinea* (Duke University Press, 2019).

8. Anna Lowenhaupt Tsing, *The Mushroom at the End of the World: On the Possibility of Life in Capitalist Ruins* (Princeton University Press, 2015); Donna J. Haraway, "Anthropocene, Capitalocene, Plantationocene, Chthulucene: Making Kin," *Environmental Humanities* 6, no. 1 (2015): 159–65; Kathryn Yusoff, "Politics of the Anthropocene: Formation of the Commons as a Geologic Process," *Antipode* 50, no. 1 (2017): 255–76.
9. Marie D. Jackson et al., "Phillipsite and Al-Tobermorite Mineral Cements Produced through Low-Temperature Water-Rock Reactions in Roman Marine Concrete," *American Mineralogist* 102, no. 7 (2017): 1435–50.
10. Ravic Nijbroek, "Mangroves, Mudbanks and Seawalls: Political Ecology of Adaptation to Sea Level Rise in Suriname" (PhD diss., University of South Florida, 2012); Sarah E. Vaughn, "Erosion by Design: Rethinking Innovation, Sea Defense, and Credibility in Guyana," *Comparative Studies in Society and History* 64, no. 4 (2022): 849–77; Yusoff, "Politics of the Anthropocene."
11. Sarah E. Vaughn, "Disappearing Mangroves: The Epistemic Politics of Climate Adaptation in Guyana," *Cultural Anthropology* 32, no. 2 (2017): 242–68.
12. Marcel Vellinga, "The Inventiveness of Tradition: Vernacular Architecture and the Future," *Perspectives in Vernacular Architecture* 13, no. 2 (2006): 115–28; "Vernacular Architecture and Climate Change," Oxford Urbanists, January 10, 2020, https://www.oxfordurbanists.com/oxford-urbanists-monthly/2020/1/5/vernacular-architecture-and-climate-change.
13. John C. Everitt, "The Growth and Development of Belize City," *Journal of Latin American Studies* 18, no. 1 (1986): 75–111; Elizabeth Pigou-Dennis and Adam Grydehoj, "Accidental and Ideal Island Cities; Islanding Processes and Urban Design in Belize City and the Urban Archipelagos of Europe," *Island Studies Journal* 9, no. 2 (2014): 259–76.
14. Bonnie J. McCay, "The Littoral and the Liminal: Challenges to the Management of the Coastal and Marine Commons," *MAST* 7, no. 1 (2008): 8–9.
15. Antonio Gramsci, *Selections from the Prison Notebooks of Antonio Gramsci*. (International Publishers, 1972).

16. From McCay 2008: "As the geographer Philip Steinberg (2001) has argued, territorialization of the oceans has been a halting and incomplete process, with much of the ocean still conceptualized essentially as 'anti-civilization,' a deep void."
17. Alex Golub, "Sustainable Development as Cargo Cult: Strange Tales of Scale Making from Melanesia and Beyond" (American Anthropological Association, 2007).
18. McCay, "The Littoral and the Liminal," 8 David M. Bush, Orrin H. Pilkey, and William J. Neal, *Living by the Rules of the Sea* (Duke University Press, 1996).
19. Karl Polanyi, *The Great Transformation: The Political and Economic Origins of Our Time* (Beacon, 1944).
20. Iain Thomson, "Heidegger's Aesthetics," in *The Stanford Encyclopedia of Philosophy*, ed. Edward N. Zalta, Fall 2015, http://plato.stanford.edu/archives/fall2015/entries/heidegger-aesthetics/.
21. Thomson, "Heidegger's Aesthetics."
22. Marshall Sahlins, *Islands of History* (University of Chicago Press, 1985), 138.
23. Richard L. Woods, S. Tomlinson-Reid, and Alastair M. Reid, *The Field Guide to Ambergris Caye, Belize, C. A., Including Other Atolls* (Sue Tomlinson Guidebooks, 1988).
24. Erik Paredis et al., *The Concept of Ecological Debt: Its Meaning and Applicability in International Policy* (Academia, 2009), 40.
25. "As far as Honduras is concerned, Mitch has put an end to the twentieth century," said the Honduran president Carlos Flores. Cited at https://solidarity-us.org/node/1752, accessed August 24, 2017.
26. Jenny Pearce, "Natural and Unnatural Disasters in Central America," in *South America, Central America and the Caribbean 2003*, ed. Jacqueline West (Taylor & Francis, 2002).
27. Patrick Bond, *Looting Africa: The Economics of Exploitation* (Zed, 2006); Patrick Bond, "Repaying Africa for Climate Crisis: 'Ecological Debt' as a Development Finance Alternative to Emissions Trading," in *Upsetting the Offset: The Political Economy of Carbon Markets*, ed. Steffen Böhm and Siddhartha Dabhi (MayFly, 2009), 151–59.
28. U. Thara Srinivasan et al., "The Debt of Nations and the Distribution of Ecological Impacts from Human Activities," *Proceedings of the National Academy of Sciences* 105, no. 5 (2008): 1768–73.

29. Sigma Colón, "Environment in Debt," *Social Text*, September 13, 2011, http://socialtextjournal.org/periscope_article/environment_in_debt/.
30. Colón, "Environment in Debt."
31. Margaret Atwood, *Payback: Debt and the Shadow Side of Wealth* (House of Anansi, 2008).
32. Adale Ramos, "Climate Change Costing Central America US$105 Billion and Counting . . .," *Amandala Newspaper*, November 24, 2009, http://amandala.com.bz/news/climate-change-costing-central-america-us105-billion-and-counting/.
33. Current debt movements suggest the potential collective political power of debt and debt strikes. See, for instance, "Rolling Jubilee Student Debt Buy," Strike Debt, September 16, 2014, http://strikedebt.org/debtbuy4/.
34. The National Hurricane Center warned people that water could return very rapidly to the shallow bay: Jack Beven, "Tropical Cyclone Report," National Hurricane Center, January 29, 2001, http://www.nhc.noaa.gov/data/tcr/AL152000_Keith.pdf.
35. And the response was, somewhat like Bush after 9–11, to argue "Fear not, we'll be open for consumption." The island's primary newspaper led its first story after the storm, just five days after its departure, with the headline "Ambergris Caye Survives—Alive and Well After Keith," while the first line of the story read, "Never Fear—We're Still Here and We'll Be Ready for Tourist Season!!! is the message the people of Ambergris Caye want the world to know." *San Pedro Sun*, October 9, 2000, https://ambergriscaye.com/sanpedrosun/old/00-371.html. And from a blog account of the article (interestingly for purposes of this book from a geodesic dome enthusiast): "We were all in the loft part of the dome. I didn't feel a thing, no movement at all. With the windows closed, we couldn't even hear anything," said one neighbor who asked to remain anonymous since the island's department of tourism is discouraging "negative reports" (Freda Parker, "Hurricane Keith Huffed and Puffed and . . .," Monolithic Dome Institute, December 21, 20001, http://www.monolithic.org/xanadu-resort/hurricane-keith-huffed-and-puffed-and).
36. Nandini Gunewardena and Mark Schuller, eds., *Capitalizing on Catastrophe: Neoliberal Strategies in Disaster Reconstruction* (AltaMira, 2008); Anthony Oliver-Smith and Susanna M. Hoffman, eds., *The Angry*

*Earth: Disaster in Anthropological Perspective* (Routledge, 1999); Theodore Steinberg, *Acts of God: The Unnatural History of Natural Disaster in America* (Oxford University Press, 2006).

37. In the *Politics of Nature*, Bruno Latour argues that "ecological crises, as we have interpreted them, present themselves as *generalized revolts of the means:* no entity—whale, river, climate, earthworm, tree, calf, cow, pig, brood—agrees any longer to be treated 'simply as a means' but insists on being treated 'always as an end'" (*Politics of Nature: How to Bring the Sciences Into Democracy* [Harvard University Press, 2004], 155–56).
38. James Ferguson, *Give a Man a Fish: Reflections on the New Politics of Distribution* (Duke University Press, 2015), 31.
39. Robert Costanza et al., "The Value of Ecosystem Services: Putting the Issues in Perspective," *Ecological Economics* 25, no. 1 (1998): 67–72; Joshua Farley, "Ecosystem Services: The Economics Debate," *Ecosystem Services* 1, no. 1 (2012): 40–49.
40. Haraway, "Anthropocene, Capitalocene, Plantationocene, Chthulucene"; Jason W. Moore, "The Capitalocene, Part I: On the Nature and Origins of Our Ecological Crisis," *Journal of Peasant Studies* 44, no. 3 (2017): 594–630.

## POSTSCRIPT: WEEDY COASTAL COMMONS

1. "Stinking Sargassum Seaweed Smelling of Rotten Eggs Terrorizes Beaches," Mail Online, October 18, 2015, http://www.dailymail.co.uk/news/article-3277720/Stinking-sargassum-seaweed-smells-rotten-eggs-terrorizes-beaches-Caribbean-Florida-Texas-shores-experiencing-piles-kelp-10-feet-high.html.
2. Victor Smetacek and Adriana Zingone, "Green and Golden Seaweed Tides on the Rise," *Nature* 504 (2013): 84–88; Mengqiu Wang et al., "The Great Atlantic Sargassum Belt," *Science* 365, no. 6448 (2019): 83–87.
3. Vinay Gidwani, "Six Theses on Waste, Value, and Commons," *Social & Cultural Geography* 14, no. 7 (2013): 780.
4. Max Liboiron, Manuel Tironi, and Nerea Calvillo, "Toxic Politics: Acting in a Permanently Polluted World," *Social Studies of Science*, July 2, 2018, 332.

5. Ellie Irons, "Weedy Resistance: Multispecies Tactics for Contesting 'The Age of Man,'" *Inhabiting the Anthropocene* (blog), May 3, 2017, https://inhabitingtheanthropocene.com/2017/05/03/weedy-resistance-multispecies-tactics-for-contesting-the-age-of-man/; Rosita Henry et al., "Weedy Life: Coloniality, Decoloniality, and Tropicality," ed. A. Lundberg, *eTropic: Electronic Journal of Studies in the Tropics* 22, no. 1 (2023): 236–69; Mardi Reardon-Smith, "In the Weeds: Thinking Through Anthropological and Other Social Science Approaches to Invasive Plant Species and Their Control," *Environment and Planning E: Nature and Space* 8, no. 2 (2025): 660–74; Anna Tsing, "The Buck, the Bull, and the Dream of the Stag: Some Unexpected Weeds of the Anthropocene," *Suomen Antropologi: Journal of the Finnish Anthropological Society* 42, no. 1 (2017): 3–21; Anna Lowenhaupt Tsing, *The Mushroom at the End of the World: On the Possibility of Life in Capitalist Ruins* (Princeton University Press, 2015).
6. Deborah Bird Rose, *Shimmer: Flying Fox Exuberance in Worlds of Peril* (Edinburgh University Press, 2022); James Ferguson, *Presence and Social Obligation: An Essay on the Share* (Prickly Paradigm, 2021).
7. National Oceanic and Atmospheric Administration, US Department of Commerce, "What Is Sargassum?," accessed July 29, 2019, https://oceanexplorer.noaa.gov/facts/sargassum.html.
8. Marisol de la Cadena, "Uncommons," Society for Cultural Anthropology, March 29, 2018, https://culanth.org/fieldsights/uncommons.
9. July Hazard, "Waterfront Sonnets," *Community Development Journal* 54, no. 1 (2019): 18.
10. Irina Velicu and Gustavo García-López, "Thinking the Commons Through Ostrom and Butler: Boundedness and Vulnerability," *Theory, Culture & Society* 35, no. 6 (2018): 65.
11. De la Cadena, "Uncommons."
12. Sylvia Federici, "Feminism and the Politics of the Commons," *The Commoner*, 2010, http://www.commoner.org.uk/?p=113; Zofia Łapniewska, "Reading Elinor Ostrom Through a Gender Perspective," *Feminist Economics*, May 6, 2016, http://www.tandfonline.com/doi/abs/10.1080/13545701.2016.1171376.
13. Bhim Adhikari and Jon C. Lovett, "Institutions and Collective Action: Does Heterogeneity Matter in Community-Based Resource Management?," *Journal of Development Studies* 42, no. 3 (2006): 426–45; Janet

Henshall Momsen, "NGOs, Gender and Indigenous Grassroots Development," *Journal of International Development* 14, no. 6 (2002): 859–67; Anna Tsing, J. Peter Brosius, and Charles Zerner, "Introduction," in *Communities and Conservation* (Alta Mira, 2005).

14. Silvia Federici, *Caliban and the Witch: Women, the Body and Primitive Accumulation* (Autonomedia, 2004).
15. Judith Butler, *Gender Trouble: Feminism and the Subversion of Identity* (Routledge, 2006); George Caffentzis and Silvia Federici, "Commons Against and Beyond Capitalism," *Community Development Journal* 49, suppl. 1 (2014): i92–105; Melissa García Lamarca, "Insurgent Acts of Being-in-Common and Housing in Spain: Making Urban Commons?," in *Moving Beyond State and Market*, ed. Mary Dellenbaugh et al. (Birkhäuser Verlag GmbH, 2015), 165–76; Gustavo A. García López, Irina Velicu, and Giacomo D'Alisa, "Performing Counter-Hegemonic Common(s) Senses: Rearticulating Democracy, Community and Forests in Puerto Rico," *Capitalism Nature Socialism* 28, no. 3 (2017): 88–107; James McCarthy, "Commons as Counterhegemonic Projects," *Capitalism Nature Socialism* 16, no. 1 (2005): 9–24.
16. Patrick Bresnihan, "The More-than-Human Commons: From Commons to Commoning," in *Space, Power and the Commons: The Struggle for Alternative Futures* (Routledge, 2016), 96.
17. Bresnihan, "The More-than-Human Commons," 106.
18. Tsing, "The Buck, the Bull, and the Dream of the Stag."
19. Donna J. Haraway, *Staying with the Trouble: Making Kin in the Chthulucene* (Duke University Press, 2016).
20. Eben Kirksey, *Emergent Ecologies* (Duke University Press, 2015).
21. Amelia Moore, "Anthropocene Anthropology: Reconceptualizing Contemporary Global Change," *Journal of the Royal Anthropological Institute* 22, no. 1 (2016): 39.
22. George W. Stocking, "From Physics to Ethnology: Franz Boas' Arctic Expedition as a Problem in the Historiography of the Behavioral Sciences," *Journal of the History of the Behavioral Sciences* 1, no. 1 (1965): 53–66; Stefan Helmreich, "Nature/Culture/Seawater," *American Anthropologist* 113, no. 1 (2011): 132–44; Bronislaw Malinowski, *Argonauts of the Western Pacific: An Account of Native Enterprise and Adventure in the Archipelagoes of Melanesian New Guinea* (Routledge, 1922).

23. Tsing, "The Buck, the Bull, and the Dream of the Stag"; Ashley Carse, "Dirty Landscapes: How Weediness Indexes State Disinvestment and Global Disconnection," in *Infrastructure, Environment, and Life in the Anthropocene*, ed. Kregg Hetherington (Duke University Press, 2019), 97–114.
24. "The Buck, the Bull, and the Dream of the Stag," 7.
25. Bonnie McCay, "The Littoral and the Liminal: Challenges to the Management of the Coastal and Marine Commons." *MAST* 7, no. 1 (2008): 7–30.
26. James M. Acheson, "Anthropology of Fishing," *Annual Review of Anthropology* 10 (1981): 275–316; James M. Acheson, "The Lobster Fiefs: Economic and Ecological Effects of Territoriality in the Maine Lobster Industry," *Human Ecology* 3, no. 3 (1975): 183–207; McCay, "The Littoral and the Liminal"; Bonnie J. McCay, *Oyster Wars and the Public Trust: Property, Law, and Ecology in New Jersey History* (University of Arizona Press, 1998); F. Berkes et al., "The Benefits of the Commons," *Nature* 340, no. 6229 (1989): 91–93.
27. De la Cadena, "Uncommons."
28. Anne Sutherland, *The Making of Belize: Globalization in the Margins* (Bergin & Garvey, 1998).
29. Aldous Huxley, *Beyond the Mexique Bay* (Harper & Bros., 1934); Melissa A. Johnson, *Becoming Creole: Nature and Race in Belize* (Rutgers University Press, 2018); Sutherland, *The Making of Belize.*
30. Huxley, *Beyond the Mexique Bay*, 35.
31. López, Velicu, and D'Alisa, "Performing Counter-Hegemonic Common(s) Senses."
32. Tsing, "The Buck, the Bull, and the Dream of the Stag."
33. Carse, "Dirty Landscapes," 101.
34. Liboiron, Tironi, and Calvillo, "Toxic Politics."
35. Liboiron, Tironi, and Calvillo, "Toxic Politics," 332.
36. Gustavo A. García López, "Reflections on Disaster Colonialism: Response to Yarimar Bonilla's 'The Wait of Disaster,'" *Political Geography* 78 (2020): 102170.
37. Liboiron, Tironi, and Calvillo, "Toxic Politics"; Alexis Shotwell, *Against Purity: Living Ethically in Compromised Times* (University of Minnesota Press, 2016); Donna J. Haraway and Thyrza Nichols Goodeve,

"Modest_Witness@Second_Millennium," in *Modest_Witness@Second_Millennium. FemaleMan_Meets_OncoMouse*, 2nd ed. (Routledge, 2018), 37.

38. Maria Kaika, "'Don't Call Me Resilient Again!': The New Urban Agenda as Immunology . . . or . . . What Happens When Communities Refuse to Be Vaccinated with 'Smart Cities' and Indicators," *Environment and Urbanization* 29, no. 1 (2017): 89–102.
39. Moore, "Anthropocene Anthropology," 28.
40. Patrick Bresnihan, "John Clare and the Manifold Commons," *Environmental Humanities* 3, no. 1 (2013): 88.
41. Marcel Mauss, *The Gift: Forms and Functions of Exchange in Archaic Societies*, trans. Ian Cunnison (Martino, 2011).

# BIBLIOGRAPHY


Acheson, James M. "Anthropology of Fishing." *Annual Review of Anthropology* 10 (1981): 275–316.

——. "The Lobster Fiefs: Economic and Ecological Effects of Territoriality in the Maine Lobster Industry." *Human Ecology* 3, no. 3 (1975): 183–207.

Adhikari, Bhim, and Jon C. Lovett. "Institutions and Collective Action: Does Heterogeneity Matter in Community-Based Resource Management?" *Journal of Development Studies* 42, no. 3 (2006): 426–45.

Agrawal, Arun. *Environmentality: Technologies of Government and the Making of Subjects*. Duke University Press, 2005.

Aitken, Stuart C., and Mei-Po Kwan. "GIS as Qualitative Research: Knowledge, Participatory Politics and Cartographies of Affect." In *The SAGE Handbook of Qualitative Geography*, ed. Dydia DeLyser, Steve Herbert, Stuart Aitken, Mike Crang, and Linda McDowell. SAGE, 2010.

Akerman, Maria. "What Does 'Natural Capital' Do? The Role of Metaphor in Economic Understanding of the Environment." *Environmental Values* 12, no. 4 (2003): 431–48.

Alcamo, Joseph, Elena M. Bennett, Rashid Hassan, and Millennium Ecosystem Assessment (Program). *Ecosystems and Human Well-Being: A Framework for Assessment*. Island, 2003.

Anand, Nikhil. *Hydraulic City: Water and the Infrastructures of Citizenship in Mumbai*. Duke University Press, 2017.

——. "Leaky States: Water Audits, Ignorance, and the Politics of Infrastructure." *Public Culture* 27, no. 2 (76) (2015): 305–30.

——. "Pressure: The PoliTechnics of Water Supply in Mumbai." *Cultural Anthropology* 26, no. 4 (2012): 542–64.

Anand, Nikhil, Akhil Gupta, and Hannah Appel, eds. *The Promise of Infrastructure*. Duke University Press, 2018.

Anderson, Eugene Newton. *Political Ecology in a Yucatec Maya Community*. University of Arizona Press, 2005.

Anker, Peder. "The Closed World of Ecological Architecture." *Journal of Architecture* 10, no. 5 (2005): 527–52.

——. *From Bauhaus to Ecohouse: A History of Ecological Design*. Louisiana State University Press, 2010.

——. *Imperial Ecology: Environmental Order in the British Empire, 1895–1945*. Harvard University Press, 2001.

Appel, Hannah. *The Licit Life of Capitalism: U.S. Oil in Equatorial Guinea*. Duke University Press, 2019.

Appel, Hannah, and Mukul Kumar. "Finance." Society for Cultural Anthropology, September 24, 2015. https://culanth.org/fieldsights/finance.

Arendt, Hannah. *Men in Dark Times*. Mariner, 1970.

Aronson, James, Suzanne J. Milton, and James N. Blignaut. *Restoring Natural Capital: Science, Business, and Practice*. Island, 2012.

Atwood, Margaret. *Payback: Debt and the Shadow Side of Wealth*. House of Anansi Press, 2008.

Bakker, Karen. "Neoliberal Nature, Ecological Fixes, and the Pitfalls of Comparative Research." *Environment and Planning A* 41, no. 8 (2009): 1781–87.

Balam, Osmer. "Overt Language Attitudes and Linguistic Identities Among Multilingual Speakers in Northern Belize." *Studies in Hispanic and Lusophone Linguistics* 6, no. 2 (2013): 247–77.

Baudrillard, Jean. *The Illusion of the End*. Stanford University Press, 1994.

——. *Simulacra and Simulation*. University of Michigan Press, 1994.

Bear, Laura, Karen Ho, Anna Tsing, and Sylvia Yanagisako. "Gens: A Feminist Manifesto for the Study of Capitalism." Society for Cultural Anthropology, March 30, 2015. https://culanth.org/fieldsights/gens-a-feminist-manifesto-for-the-study-of-capitalism.

Becker, Egon, and Thomas Jahn. "Growth or Development?" In *Political Ecology: Global and Local*, ed. Roger Keil. Routledge, 1998.

Benton, Adia. "African Expatriates and Race in the Anthropology of Humanitarianism." *Critical African Studies* 8, no. 3 (2016): 266–77.

Berkes, F., D. Feeny, B. J. McCay, and J. M. Acheson. "The Benefits of the Commons." *Nature* 340, no. 6229 (1989): 91–93.

Besky, Sarah, and Alex Blanchette, eds. *How Nature Works: Rethinking Labor on a Troubled Planet*. University of New Mexico Press published in association with the School for Advanced Research Press, 2019.

Bessire, Lucas, and David Bond. "Ontological Anthropology and the Deferral of Critique." *American Ethnologist* 41, no. 3 (2014): 440–56.

Blaser, Mario. "Political Ontology: Cultural Studies Without 'Cultures'?" *Cultural Studies* 23, no. 5–6 (2009): 873–96.

Bocking, Stephen. *Ecologists and Environmental Politics: A History of Contemporary Ecology*. Yale University Press, 1997.

——. "The Ecosystem: Research and Practice in North America." *Web Ecology* 13 (2013): 43–47.

——. "Ecosystems, Ecologists, and the Atom: Environmental Research at Oak Ridge National Laboratory." *Journal of the History of Biology* 28, no. 1 (1995): 1–47.

Bolland, O. Nigel. *Colonialism and Resistance in Belize: Essays in Historical Sociology*. Cubola Productions, 2003.

——. *The Formation of a Colonial Society: Belize, from Conquest to Crown Colony*. Johns Hopkins University Press, 1977.

Bond, Patrick. *Looting Africa: The Economics of Exploitation*. Zed, 2006.

——. "Repaying Africa for Climate Crisis: 'Ecological Debt' as a Development Finance Alternative to Emissions Trading." In *Upsetting the Offset: The Political Economy of Carbon Markets*, ed. Steffen Böhm and Siddhartha Dabhi. MayFly, 2009.

Bonilla, Yarimar. "Unsettling Sovereignty." *Cultural Anthropology* 32, no. 3 (2017): 330–39.

Boulding, Kenneth. "The Economics of the Coming Spaceship Earth." In *Environmental Quality in a Growing Economy*, ed. Henry Jarrett. Resources for the Future Press, 1966.

Bowker, Geoffrey C. "Time, Money, and Biodiversity." In *Global Assemblages*, ed. Aihwa Ong and Stephen J. Collier. Blackwell, 2005.

Braat, Leon C., and Rudolf de Groot. "The Ecosystem Services Agenda: Bridging the Worlds of Natural Science and Economics, Conservation and Development, and Public and Private Policy." *Ecosystem Services* 1, no. 1 (2012): 4–15.

Brand, Stewart. "We Are as Gods." In *Whole Earth Catalog*. Portola Institute, 1968.

Bresnihan, Patrick. "John Clare and the Manifold Commons." *Environmental Humanities* 3, no. 1 (2013): 71–91.

——. "The More-than-Human Commons: From Commons to Commoning." In *Space, Power and the Commons: The Struggle for Alternative Futures*. Routledge, 2016.

Brockington, Dan D. *Fortress Conservation: The Preservation of the Mkomazi Game Reserve, Tanzania*. Indiana University Press, 2002.

Büscher, Bram. "Anti-Politics as Political Strategy: Neoliberalism and Transfrontier Conservation in Southern Africa." *Development and Change* 41, no. 1 (2010): 29–51.

Bush, David M., Orrin H. Pilkey, and William J. Neal. *Living by the Rules of the Sea*. Duke University Press, 1996.

Butler, Judith. *Gender Trouble: Feminism and the Subversion of Identity*. Routledge, 2006.

Caffentzis, George, and Silvia Federici. "Commons Against and Beyond Capitalism." *Community Development Journal* 49, suppl. 1 (2014): i92–i105.

Callon, Michel. "Civilizing Markets: Carbon Trading Between *In Vitro* and *In Vivo* Experiments." *Accounting, Organizations and Society* 34, no. 3–4 (2009): 535–48.

——. "Some Elements of a Sociology of Translation: Domestication of the Scallops and the Fishermen of Saint Brieuc Bay." In *The Science Studies Reader*. Routledge, 1999.

Camargo, Alejandro. "On the Translucence of Water." *Engagement (Anthropology and Environment)* (blog), February 8, 2024. https://aesengagement.wordpress.com/2024/02/07/on-the-translucence-of-water/.

Carrier, James G., and Daniel Miller. *Virtualism: A New Political Economy*. Berg, 1998.

Carrier, James G., and Paige West. *Virtualism, Governance and Practice*. Berghahn, 2009.

Carse, Ashley. "Dirty Landscapes: How Weediness Indexes State Disinvestment and Global Disconnection." In *Infrastructure, Environment, and Life in the Anthropocene*, ed. Kregg Hetherington. Duke University Press, 2019.

——. "Nature as Infrastructure: Making and Managing the Panama Canal Watershed." *Social Studies of Science* 42, no. 4 (2012): 539–63.

Cassidy, W. B., ed. *Bioengineering and Cabin Ecology*. Vol. 20 of the Science and Technology series. American Astronautical Society, 1969.

Castree, Noel. "Neoliberalising Nature: The Logics of Deregulation and Reregulation." *Environment and Planning A* 40, no. 1 (2008): 131–52.

——. "Neoliberalism and the Biophysical Environment 1: What 'Neoliberalism' Is, and What Difference Nature Makes to It." *Geography Compass* 4, no. 12 (2010): 1725–33.

Chakrabarty, Dipesh. *Rethinking Working-Class History: Bengal, 1890–1940*. Princeton University Press, 2000.

Choy, Timothy. *Ecologies of Comparison: An Ethnography of Endangerment in Hong Kong*. Duke University Press, 2011.

Clifford, James, and George E. Marcus. *Writing Culture: The Poetics and Politics of Ethnography*. University of California Press, 1986.

Clough, Patricia Ticineto, and Jean Halley, eds. *The Affective Turn: Theorizing the Social*. Duke University Press, 2007.

Coetzee, Serena, Ivana Ivánová, Helena Mitasova, and Maria Antonia Brovelli. "Open Geospatial Software and Data: A Review of the Current State and A Perspective into the Future." *ISPRS International Journal of Geo-Information* 9, no. 2 (2020): 90.

Cohen, Erik. "Expatriate Communities." *Current Sociology* 24, no. 3 (1977): 5–90.

Collier, Stephen J. "The Spatial Forms and Social Norms of 'Actually Existing Neoliberalism': Toward a Substantive Analytics." Working paper in *International Affairs*. The New School, Graduate Program in International Affairs, 2005.

——. *Post-Soviet Social: Neoliberalism, Social Modernity, Biopolitics*. Princeton University Press, 2011.

Colón, Sigma. "Environment in Debt." *Social Text*, September 13, 2011. http://socialtextjournal.org/periscope_article/environment_in_debt/.

"Confusion Matrix—GIS Wiki | The GIS Encyclopedia." Accessed January 14, 2019. http://wiki.gis.com/wiki/index.php/Confusion_matrix.

Cooper, Emily, Lauretta Burke, and Nadia Bood. "Coastal Capital: Belize. The Economic Contribution of Belize's Coral Reefs and Mangroves." WRI Working Paper. World Resources Institute, 2009. http://www.wri.org/publications.

Cortesi, Luisa. "An Ontology of Water and Land in North Bihar, India★." *Journal of the Royal Anthropological Institute* 27, no. 4 (2021): 870–89.

Costanza, Robert. "Valuing Natural Capital and Ecosystem Services Toward the Goals of Efficiency, Fairness, and Sustainability." *Ecosystem Services* 43 (2020): 101096.

Costanza, Robert, Ralph d'Arge, Rudolf de Groot, Stephen Farber, Monica Grasso, et al. "The Value of the World's Ecosystem Services and Natural Capital." *Nature* 387, no. 6630 (1997): 253–60.

Costanza, Robert, Jose Paruelo, Robert G. Raskin, Paul Sutton, Marjan van den Belt, et al. "The Value of Ecosystem Services: Putting the Issues in Perspective." *Ecological Economics* 25, no. 1 (1998): 67–72.

Cox, Courtney E., Corbin D. Jones, John P. Wares, Karl D. Castillo, Melanie D. McField, et al. "Genetic Testing Reveals Some Mislabeling but General Compliance with a Ban on Herbivorous Fish Harvesting in Belize." *Conservation Letters* 6, no. 2 (2013): 132–40.

Crona, Beatrice, Magnus Nyström, Carl Folke, and Narriman Jiddawi. "Middlemen, a Critical Social-Ecological Link in Coastal Communities of Kenya and Zanzibar." *Marine Policy* 34, no. 4 (2010): 761–71.

Daggett, Cara. "Petro-Masculinity: Fossil Fuels and Authoritarian Desire." *Millennium* 47, no. 1 (2018): 25–44.

Daily, Gretchen C. *Nature's Services*. Island, 1997.

Daily, Gretchen C., Tore Söderqvist, Sara Aniyar, Kenneth Arrow, Partha Dasgupta, et al. "The Value of Nature and the Nature of Value." *Science* 289, no. 5478 (2000): 395–96.

Daly, Herman E., John B. Cobb, and Clifford W. Cobb. *For the Common Good*. Beacon, 1994.

Danely, Jason. "Affect, Infrastructure, and Vulnerability: Making and Breaking Japanese Eldercare." *Medicine Anthropology Theory* 3, no. 2 (2016): 198.

Davis, Heather, and Zoe Todd. "On the Importance of a Date, or, Decolonizing the Anthropocene." *ACME: An International Journal for Critical Geographies* 16, no. 4 (2017): 761–80.

Debord, Guy. *The Society of the Spectacle*. Zone, 1994.

De Castro, Eduardo Viveiros. "Cosmological Deixis and Amerindian Perspectivism." *Journal of the Royal Anthropological Institute* 4, no. 3 (1998): 469–88.

——. "On Models and Examples: Engineers and Bricoleurs in the Anthropocene." *Current Anthropology* 60, no. S20 (2019): S296–308.

De la Cadena, Marisol. "Uncommons." Society for Cultural Anthropology, March 29, 2018. https://culanth.org/fieldsights/uncommons.

De la Cadena, Marisol, Robert J. Foster, and Daniel R. Reichman. *Earth Beings: Ecologies of Practice Across Andean Worlds*. Illus. ed. Duke University Press Books, 2015.

Dempsey, Caitlin. "GIS Job Definitions." *GIS Lounge* (blog), April 17, 2022. https://www.gislounge.com/gis-job-definitions/.

Dempsey, Jessica, and Morgan M. Robertson. "Ecosystem Services Tensions, Impurities, and Points of Engagement within Neoliberalism." *Progress in Human Geography* 36, no. 6 (2012): 758–79.

Dempsey Morais, Caitlyn. "The GIS Career, Revisioned." *GIS Lounge* (blog), 2012. http://www.gislounge.com/the-gis-career-revisioned/.

DePuy, Walker, Jacob Weger, Katie Foster, Anya M Bonanno, Suneel Kumar, Kristen Lear, Raul Basilio, and Laura German. "Environmental Governance: Broadening Ontological Spaces for a More Livable World." *Environment and Planning E: Nature and Space* 5, no. 2 (2022): 947–75.

Derrida, Jacques. *Writing and Difference*. University of Chicago Press, 1978.

Devji, Faisal. *The Terrorist in Search of Humanity: Militant Islam and Global Politics*. Columbia University Press, 2008.

DiBiase, David. "Strengthening the GIS Profession." *ArcNews*, 2012. http://www.esri.com/news/arcnews/summer12articles/strengthening-the-gis-profession.html.

Dove, Michael R. "The Dialectical History of 'Jungle' in Pakistan: An Examination of the Relationship Between Nature and Culture." SSRN Scholarly Paper. Social Science Research Network, 1992. http://papers.ssrn.com/abstract=2403490.

Dutt, Rajeshwari. *Empire on Edge: The British Struggle for Order in Belize During Yucatan's* Caste *War, 1847–1901*. Cambridge University Press, 2020.

Edwards, Paul N. *The Closed World: Computers and the Politics of Discourse in Cold War America*. MIT Press, 1997.

——. *A Vast Machine: Computer Models, Climate Data, and the Politics of Global Warming*. MIT Press, 2010.

Ehrlich, Paul. *The Dominant Animal: Human Evolution and the Environment*. Island, 2008.

——. *The Population Bomb*. Ballantine, 1968.

Ehrlich, Paul R., and Anne H. Ehrlich. *Extinction: The Causes and Consequences of the Disappearance of Species*. Random House, 1981.

Ehrlich, Paul R., and Peter H. Raven. "Butterflies and Plants: A Study in Coevolution." *Evolution* 18, no. 4 (1964): 586–608.

Elizabeth Pigou-Dennis and Adam Grydehoj. "Accidental and Ideal Island Cities; Islanding Processes and Urban Design in Belize City and the Urban Archipelagos of Europe." *Island Studies Journal* 9, no. 2 (2014): 259–76.

Elton, Charles S. *Animal Ecology*. University of Chicago Press, 1927.

Everitt, John C. "The Growth and Development of Belize City." *Journal of Latin American Studies* 18, no. 1 (1986): 75–111.

——. "The Torch Is Passed: Neocolonialism in Belize." *Caribbean Quarterly* 33, no. 3/4 (1987): 42–59.

Farley, Joshua. "Ecosystem Services: The Economics Debate." *Ecosystem Services* 1, no. 1 (2012): 40–49.

Federici, Silvia. *Caliban and the Witch: Women, the Body and Primitive Accumulation*. Autonomedia, 2004.

Federici, Sylvia. "Feminism and the Politics of the Commons." *The Commoner*, 2010. http://www.commoner.org.uk/?p=113.

Fenichel, Eli P., and Joshua K. Abbott. "Natural Capital: From Metaphor to Measurement." *Journal of the Association of Environmental and Resource Economists* 1, no. 1/2 (2014): 1–27.

Ferdinand, Malcom. *Decolonial Ecology: Thinking from the Caribbean World*. Trans. Anthony Paul Smith. Polity, 2022.

Ferguson, James. *Give a Man a Fish: Reflections on the New Politics of Distribution*. Duke University Press, 2015.

——. *Presence and Social Obligation: An Essay on the Share*. Prickly Paradigm, 2021.

——. "Seeing Like an Oil Company: Space, Security, and Global Capital in Neoliberal Africa." *American Anthropologist* 107, no. 3 (2005): 377–82.

——. "Toward a Left Art of Government: From 'Foucauldian Critique' to Foucauldian Politics." *History of the Human Sciences* 24, no. 4 (2011): 61–68.

——. "The Uses of Neoliberalism." *Antipode* 41, suppl. 1 (2010): 166–84.

Fletcher, Robert. *Failing Forward: The Rise and Fall of Neoliberal Conservation*. University of California Press, 2023.

——. "How I Learned to Stop Worrying and Love the Market: Virtualism, Disavowal, and Public Secrecy in Neoliberal Environmental Conservation." *Environment and Planning D: Society and Space* 31, no. 5 (2013): 796.

——. "Neoliberal Environmentality: Toward a Poststructuralist Political Ecology of the Conservation Debate." *Conservation and Society* 8, no. 3 (2010): 171.

Fletcher, Robert, and Jan Breitling. "Market Mechanism or Subsidy in Disguise? Governing Payment for Environmental Services in Costa Rica." *Geoforum* 43, no. 3 (2012): 402–11.

Foody, Giles M. "Status of Land Cover Classification Accuracy Assessment." *Remote Sensing of Environment* 80, no. 1 (2002): 185–201.

Frenay, Robert. *Pulse: The Coming Age of Systems and Machines Inspired by Living Things*. Macmillan, 2006.

Fuller, Richard Buckminster. *Operating Manual for Spaceship Earth*. Simon and Schuster, 1970.

Galison, Peter, and Bruce William Hevly. *Big Science: The Growth of Large-Scale Research*. Stanford University Press, 1992.

Gallagher, Patrick, and Danielle DiNovelli-Lang. "Nature and Knowledge." *Environment and Society* 5, no. 1 (2014): 1–6.

García López, Gustavo A. "Reflections on Disaster Colonialism: Response to Yarimar Bonilla's 'The Wait of Disaster.'" *Political Geography* 78 (2020): 102170.

Geertz, Clifford. "Deep Play: Notes on the Balinese Cockfight." *Daedalus* 101, no. 1 (1972): 1–37.

George, Henry. *Progress and Poverty: An Inquiry Into the Cause of Industrial Depressions and of Increase of Want with Increase of Wealth; the Remedy*. National Single Tax League, 1879.

Gibson-Graham, J. K. *The End of Capitalism (As We Knew It): A Feminist Critique of Political Economy*. University of Minnesota Press, 2006.

Gidwani, Vinay. "Six Theses on Waste, Value, and Commons." *Social & Cultural Geography* 14, no. 7 (2013): 773–83.

Glissant, Édouard. *Poetics of Relation*. Trans. Betsy Wing. University of Michigan Press, 1997.

Golley, Frank Benjamin. *A History of the Ecosystem Concept in Ecology: More Than the Sum of the Parts*. Yale University Press, 1996.

Golub, Alex. "Sustainable Development as Cargo Cult: Strange Tales of Scale Making from Melanesia and Beyond." American Anthropological Association, 2007.

Gómez-Baggethun, Erik, Rudolf de Groot, Pedro L. Lomas, and Carlos Montes. "The History of Ecosystem Services in Economic Theory and Practice: From Early Notions to Markets and Payment Schemes." *Ecological Economics* 69, no. 6 (2010): 1209–18.

González-Hidalgo, Marien, and Christos Zografos. "Emotions, Power, and Environmental Conflict: Expanding the 'Emotional Turn' in Political Ecology." *Progress in Human Geography* 44, no. 2 (2020): 235–55.

GPS.gov. "The Global Positioning System." Accessed June 21, 2015. http://www.gps.gov/systems/gps/.

Graeber, David. *The Democracy Project: A History, a Crisis, a Movement*. Penguin, 2013.

——. "Radical Alterity Is Just Another Way of Saying 'Reality': A Reply to Eduardo Viveiros de Castro." *HAU: Journal of Ethnographic Theory* 5, no. 2 (2015): 1–41.

Gramsci, Antonio. *Selections from the Prison Notebooks of Antonio Gramsci*. International Publishers, 1972.

Greenleaf, Maron. "Distributive Politics of REDD in Acre, Brazil." *Anthropology News* 55, no. 6 (2014).

Greenleaf, Maron E. *Forest Lost: Producing Green Capitalism in the Brazilian Amazon*. Duke University Press, 2024.

Grove, Richard H. *Green Imperialism: Colonial Expansion, Tropical Island Edens and the Origins of Environmentalism, 1600–1860*. Cambridge University Press, 1996.

Grugel, Jean, and Pía Riggirozzi. "Post-Neoliberalism in Latin America: Rebuilding and Reclaiming the State after Crisis." *Development and Change* 43, no. 1 (2012): 1–21.

Gunewardena, Nandini, and Mark Schuller, eds. *Capitalizing on Catastrophe: Neoliberal Strategies in Disaster Reconstruction*. AltaMira, 2008.

Gusterson, Hugh. *Nuclear Rites: A Weapons Laboratory at the End of the Cold War*. University of California Press, 1998.

——. *People of the Bomb: Portraits of America's Nuclear Complex*. University of Minnesota Press, 2004.

Hagen, Joel Bartholemew. *An Entangled Bank: The Origins of Ecosystem Ecology*. Rutgers University Press, 1992.

Hamblin, Jacob Darwin. *Arming Mother Nature: The Birth of Catastrophic Environmentalism*. Oxford University Press, 2013.

Hammeke, Erin. "Logwood Dye on Paper," 2004. https://sentra.ischool.utexas.edu/~cochinea/pdfs/e-hammeke-04-logwood.pdf.

Haraway, Donna J. "Anthropocene, Capitalocene, Plantationocene, Chthulucene: Making Kin." *Environmental Humanities* 6, no. 1 (2015): 159–65.

——. *Staying with the Trouble: Making Kin in the Chthulucene.* Duke University Press Books, 2016.

Haraway, Donna J., and Thyrza Nichols Goodeve. "Modest_Witness@Second_Millennium." In *Modest_Witness@Second_Millennium. FemaleMan_Meets_OncoMouse*, 2nd ed. Routledge, 2018.

Hartigan, John. *Aesop's Anthropology: A Multispecies Approach.* University of Minnesota Press, 2014.

Hazard, July. "Waterfront Sonnets." *Community Development Journal* 54, no. 1 (2019): 17–21.

Hazen, Helen D., and Leila M. Harris. "Limits of Territorially-Focused Conservation: A Critical Assessment Based on Cartographic and Geographic Approaches." *Environmental Conservation* 34, no. 4 (2007): 280–90.

Healy, Hali, Sylvia Lorek, and Rodrígues-Labajos. "Ecosystem Services." EJOLT (Environmental Justice Organisations, Liabilities and Trade). Accessed July 30, 2025. http://www.ejolt.org/2013/05/ecosystem-services/.

Helmreich, Stefan. "From Spaceship Earth to Google Ocean: Planetary Icons, Indexes, and Infrastructures." *Social Research: An International Quarterly* 78, no. 4 (2011): 1211–42.

——. "Nature/Culture/Seawater." *American Anthropologist* 113, no. 1 (2011): 132–44.

Henry, Rosita, Helen Ramoutsaki, Debbi Long, Greg Acciaioli, Simon Foale, Celmara Pocock, Kristin McBain-Rigg, and Michael Wood. "Weedy Life: Coloniality, Decoloniality, and Tropicality." Ed. A. Lundberg. *eTropic: Electronic Journal of Studies in the Tropics* 22, no. 1 (2023): 236–69.

Ho, Karen. *Liquidated: An Ethnography of Wall Street.* Duke University Press, 2009.

Holbraad, Martin, Morten Axel Pedersen, and Eduardo Viveiros de Castro. "The Politics of Ontology: Anthropological Positions." Society for Cultural Anthropology, January 13, 2014. https://culanth.org/fieldsights/the-politics-of-ontology-anthropological-positions.

Hong, Jung Eun. "Identifying Skill Requirements for GIS Positions: A Content Analysis of Job Advertisements." *Journal of Geography* 115, no. 4 (2016): 147–58.

Huxley, Aldous. *Beyond the Mexique Bay.* Harper & Bros., 1934.

Igoe, James, and Dan Brockington. "Neoliberal Conservation: A Brief Introduction." *Conservation and Society* 5, no. 4 (2007): 232–49.

Igoe, Jim, Katja Neves, and Dan Brockington. "A Spectacular Eco-Tour Around the Historic Bloc: Theorising the Convergence of Biodiversity Conservation and Capitalist Expansion." *Antipode* 42, no. 3 (2010): 486–512.

Ingold, Tim. "Globes and Spheres: The Topology of Environmentalism." In *Environmental Anthropology: A Historical Reader*, ed. Michael Dove and Carol Carpenter. Blackwell, 2007.

Irons, Ellie. "Weedy Resistance: Multispecies Tactics for Contesting 'The Age of Man.'" *Inhabiting the Anthropocene* (blog), May 3, 2017. https://inhabitingtheanthropocene.com/2017/05/03/weedy-resistance-multispecies-tactics-for-contesting-the-age-of-man/.

Jacka, Jerry. "Place, Time, and Affect: Changing Landscapes Around a New Guinea Mining Area." *Cultural Anthropology* 37, no. 3 (2022): 549–71.

Jackson, Marie D., Sean R. Mulcahy, Heng Chen, Yao Li, Qinfei Li, Piergiulio Cappelletti, and Hans-Rudolf Wenk. "Phillipsite and Al-Tobermorite Mineral Cements Produced through Low-Temperature Water-Rock Reactions in Roman Marine Concrete." *American Mineralogist* 102, no. 7 (2017): 1435–50.

Jackson, Steven J., Paul N. Edwards, Geoffrey C. Bowker, and Cory P. Knobel. "Understanding Infrastructure: History, Heuristics and Cyberinfrastructure Policy." *First Monday* 12, no. 6 (2007).

Jasanoff, Sheila. "Image and Imagination: The Formation of Global Environmental Consciousness." In *Changing the Atmosphere: Expert Knowledge and Environmental Governance*. MIT Press, 2001.

——. *States of Knowledge: The Co-production of Science and Social Order*. Routledge, 2004.

Jervis, John. "The Nature of It All (Modernist Ontology)." In *Modernity Theory: Modern Experience, Modernist Consciousness, Reflexive Thinking*, ed. John Jervis. Palgrave Macmillan, 2018.

Johnson, Melissa A. *Becoming Creole: Nature and Race in Belize*. Rutgers University Press, 2018.

——. "The Making of Race and Place in Nineteenth-Century British Honduras." *Environmental History* 8, no. 4 (2003): 598–617.

——. "Racing Nature and Naturalizing Race: Rethinking the Nature of Creole and Garifuna Identities." *Belizean Studies* 27, no. 2 (2005): 43–56.

Kaika, Maria. "'Don't Call Me Resilient Again!': The New Urban Agenda as Immunology . . . or . . . What Happens When Communities Refuse to Be Vaccinated with 'Smart Cities' and Indicators." *Environment and Urbanization* 29, no. 1 (2017): 89–102.

Kareiva, Peter, Heather Tallis, Taylor H. Ricketts, Gretchen C. Daily, and Stephen Polasky. *Natural Capital: Theory and Practice of Mapping Ecosystem Services*. Oxford University Press, 2011.

Keane, Webb. "Ontologies, Anthropologists, and Ethical Life." *HAU: Journal of Ethnographic Theory* 3, no. 1 (2013): 186–91.

Kinzig, A. P., C. Perrings, F. S. Chapin, S. Polasky, V. K. Smith, D. Tilman, and B. L. Turner. "Paying for Ecosystem Services—Promise and Peril." *Science* 334, no. 6056 (2011): 603–4.

Kirksey, Eben. *Emergent Ecologies*. Duke University Press, 2015.

Kleypas, Joan A., and Chris Langdon. "Coral Reefs and Changing Seawater Carbonate Chemistry." In *Coral Reefs and Climate Change: Science and Management*, ed. Jonathan T. Phinney, Ove Hoegh-Guldberg, Joanie Kleypas, William Skirving, and Al Strong. American Geophysical Union, 2006.

Kolinjivadi, Vijay, Gert Van Hecken, Diana Vela Almeida, Jérôme Dupras, and Nicolás Kosoy. "Neoliberal Performatives and the 'Making' of Payments for Ecosystem Services (PES)." *Progress in Human Geography* 43, no. 1 (2019): 3–25.

Kongsager, Rico, and Esteve Corbera. "Linking Mitigation and Adaptation in Carbon Forestry Projects: Evidence from Belize." *World Development* 76 (2015): 132–46.

Kosek, Jake. "Ecologies of Empire: On the New Uses of the Honeybee." *Cultural Anthropology* 25, no. 4 (2010): 650–78.

——. *Understories: The Political Life of Forests in Northern New Mexico*. Duke University Press, 2006.

Kramer, Philip A., and Patricia R. Kramer. "Ecoregional Conservation Planning for the Mesoamerican Caribbean Reef." World Wildlife Fund, 2002.

Krueger, Rob, and David Gibbs. *The Sustainable Development Paradox: Urban Political Economy in the United States and Europe*. Guilford, 2007.

Kwa, Chunglin. "Radiation Ecology, Systems Ecology and the Management of the Environment." In *Science and Nature: Essays in the History of the Environmental Sciences*, vol. 8 (British Society for the History of Science, 1993). Lahsen, Myanna. "Seductive Simulations? Uncertainty Distribution Around Climate Models." *Social Studies of Science* 35, no. 6 (2005): 895–922.

Lamarca, Melissa García. "Insurgent Acts of Being-in-Common and Housing in Spain: Making Urban Commons?" In *Moving Beyond State and Market*, ed. Mary Dellenbaugh, Markus Kip, Agnes Katharina Müller, and Martin Schwegmann. Birkhäuser Verlag GmbH, 2015.

Land, Global, Alan H. Strahler, Luigi Boschetti, Giles M. Foody, Mark A. Friedl, et al. "Global Land Cover Validation: Recommendations for Evaluation and Accuracy Assessment of Global Land Cover Maps." Luxembourg: Office for Official Publications of the European Communities, 2006. http://landval.gsfc.nasa.gov/pdf/GlobalLandCoverValidation.pdf.

Łapniewska, Zofia. "Reading Elinor Ostrom Through a Gender Perspective." *Feminist Economics*, May 6, 2016. http://www.tandfonline.com/doi/abs/10.1080/13545701.2016.1171376.

Larkin, Brian. "The Politics and Poetics of Infrastructure." *Annual Review of Anthropology* 42, no. 1 (2013): 327–43.

Latour, Bruno. *An Inquiry Into Modes of Existence: An Anthropology of the Moderns*. Trans. Catherine Porter. Harvard University Press, 2013.

——. *Politics of Nature: How to Bring the Sciences Into Democracy*. Harvard University Press, 2004.

——. "Postmodern? No, Simply Amodern! Steps Toward an Anthropology of Science." *Studies in History and Philosophy of Science Part A* 21, no. 1 (1990): 145–71.

——. *We Have Never Been Modern*. Trans. Catherine Porter. Harvard University Press, 1993.

Leach, Edmund. "Structuralism in Social Anthropology." In *Structuralism: An Introduction*, ed. David Robey. Clarendon, 1973.

Lefebvre, Henri. *The Production of Space*. Wiley-Blackwell, 1991.

Leitner, Helga, Jamie Peck, and Eric S. Sheppard. *Contesting Neoliberalism: Urban Frontiers*. Guilford, 2007.

Leslie, Stuart W. *The Cold War and American Science: The Military-Industrial-Academic Complex at MIT and Stanford*. Columbia University Press, 1993.

Lévi-Strauss, Claude. *The Elementary Structures of Kinship*. Beacon, 1969.

——. *Structural Anthropology*. Basic Books, 1963.

Lewis, Sara M. "The Role of Herbivorous Fishes in the Organization of a Caribbean Reef Community." *Ecological Monographs* 56, no. 3 (1986): 183–200.

Li, Tania Murray. *Land's End: Capitalist Relations on an Indigenous Frontier*. Duke University Press, 2014.

Liboiron, Max, Manuel Tironi, and Nerea Calvillo. "Toxic Politics: Acting in a Permanently Polluted World:" *Social Studies of Science*, July 2, 2018.

Liew, S.C. "Optical Remote Sensing (Tutorial)." Center for Remote Image Sensing and Processing (CRISP), 2001. http://www.crisp.nus.edu.sg/~research/tutorial/optical.htm#spectral_signature.

Little, Kenneth. "Blue." *Imaginativeethnography.Org* (blog), 2014. http://imaginativeethnography.org/imaginings/affect/blue/.

Liverman, Diana M., and Rosa Maria Roman Cuesta. "Human Interactions with the Earth System: People and Pixels Revisited." *Earth Surface Processes and Landforms* 33, no. 9 (2008): 1458–71.

López, Gustavo A. García, Irina Velicu, and Giacomo D'Alisa. "Performing Counter-Hegemonic Common(s) Senses: Rearticulating Democracy, Community and Forests in Puerto Rico." *Capitalism Nature Socialism* 28, no. 3 (2017): 88–107.

Lowen, Rebecca S. *Creating the Cold War University: The Transformation of Stanford*. University of California Press, 1997.

MacCormack, Carol P. "Nature, Culture and Gender: A Critique." In *Nature, Culture and Gender*, ed. Carol P. MacCormack and Marilyn Strathern. Cambridge University Press, 1980.

Mackenzie, Donald. "Is Economics Performative? Option Theory and the Construction of Derivatives Markets." *Journal of the History of Economic Thought* 28, no. 1 (2006): 29–55.

MacKenzie, Donald. "Making Things the Same: Gases, Emission Rights and the Politics of Carbon Markets." *Accounting, Organizations and Society* 34, no. 3–4 (2009): 440–55.

Mail Online. "Stinking Sargassum Seaweed Smelling of Rotten Eggs Terrorizes Beaches," October 18, 2015. http://www.dailymail.co.uk/news/article-3277720/Stinking-sargassum-seaweed-smells-rotten-eggs-terrorizes-beaches-Caribbean-Florida-Texas-shores-experiencing-piles-kelp-10-feet-high.html.

Malinowski, Bronislaw. *Argonauts of the Western Pacific: An Account of Native Enterprise and Adventure in the Archipelagoes of Melanesian New Guinea*. Routledge, 1922.

Marsh, George Perkins. *Man and Nature: Or, Physical Geography as Modified by Human Action*. Scribner, 1864.

Martin, Emily. "The Potentiality of Ethnography and the Limits of Affect Theory." *Current Anthropology* 54, no. S7 (2013): S149–58.

Marx, Karl. *Capital: A Critique of Political Economy*. Penguin, 1976.

——. *Wages and Capital*. Labour Literature Society, 1893.

Masco, Joseph. "The End of Ends." *Anthropological Quarterly* 85, no. 4 (2012): 1107–24.

——. *The Nuclear Borderlands: The Manhattan Project in Post-Cold War New Mexico*. Princeton University Press, 2013.

——. "'SURVIVAL IS YOUR BUSINESS': Engineering Ruins and Affect in Nuclear America." *Cultural Anthropology* 23, no. 2 (2008): 361–98.

——. *The Theater of Operations: National Security Affect from the Cold War to the War on Terror*. Duke University Press, 2014.

Mathews, Andrew S. "Anthropology and the Anthropocene: Criticisms, Experiments, and Collaborations." *Annual Review of Anthropology* 49, no. 49 (2020): 67–82.

——. *Instituting Nature: Authority, Expertise, and Power in Mexican Forests*. MIT Press, 2011.

Mauss, Marcel. *The Gift: Forms and Functions of Exchange in Archaic Societies*. Trans. Ian Cunnison. Martino, 2011.

McAfee, Kathleen. "The Contradictory Logic of Global Ecosystem Services Markets." *Development and Change* 43, no. 1 (2012): 105–31.

McAfee, Kathleen, and Elizabeth N. Shapiro. "Payments for Ecosystem Services in Mexico: Nature, Neoliberalism, Social Movements, and the State." *Annals of the Association of American Geographers* 100, no. 3 (2010): 579–99.

McCarthy, James. "Commons as Counterhegemonic Projects." *Capitalism Nature Socialism* 16, no. 1 (2005): 9–24.

McCarthy, James, and Scott Prudham. "Neoliberal Nature and the Nature of Neoliberalism." *Geoforum* 35, no. 3 (2004): 275–83.

McCauley, Douglas J. "Selling Out on Nature." *Nature* 443, no. 7107 (2006): 27–28.

McCay, Bonnie J. "The Littoral and the Liminal: Challenges to the Management of the Coastal and Marine Commons." *MAST* 7, no. 1 (2008): 7–30.

——. *Oyster Wars and the Public Trust: Property, Law, and Ecology in New Jersey History*. University of Arizona Press, 1998.

McCray, W. Patrick. *The Visioneers: How a Group of Elite Scientists Pursued Space Colonies, Nanotechnologies, and a Limitless Future*. Princeton University Press, 2012.

McElwee, Pamela D. "Payments for Environmental Services as Neoliberal Market-Based Forest Conservation in Vietnam: Panacea or Problem?" *Geoforum* 43, no. 3 (2012): 412–26.

McElwee, Pamela, Tuyen Nghiem, Hue Le, Huong Vu, and Nghi Tran. "Payments for Environmental Services and Contested Neoliberalisation in Developing Countries: A Case Study from Vietnam." *Journal of Rural Studies* 36, suppl. C (2014): 423–40.

McNamee, Gregory. *Gila: The Life and Death of an American River*, updated and expanded ed. University of New Mexico Press, 2012.

McNeill, J. R., and Corinna R. Unger. "Introduction: The Big Picture." In *Environmental Histories of the Cold War*. Publications of the German Historical Institute. Cambridge University Press, 2010.

Meadows, Donella H. *Limits to Growth*. Penguin, 1972.

Medina, Laurie Kroshus. "Defining Difference, Forging Unity: The Co-construction of Race, Ethnicity and Nation in Belize." *Ethnic and Racial Studies* 20, no. 4 (1997): 757–80.

Miller, Daniel. "Turning Callon the Right Way Up." *Economy and Society* 31, no. 2 (2002): 218–33.

Mitchell, Timothy. *Rule of Experts: Egypt, Techno-Politics, Modernity*. University of California Press, 2002.

——. "The World as Exhibition." *Comparative Studies in Society and History* 31, no. 2 (1989): 217–36.

Mody, Cyrus C. M. "How I Learned to Stop Worrying and Love the Bomb, the Nuclear Reactor, the Computer, Ham Radio, and Recombinant DNA." *Historical Studies in the Natural Sciences*38, no. 3 (2008): 451–61.

Momsen, Janet Henshall. "NGOs, Gender and Indigenous Grassroots Development." *Journal of International Development* 14, no. 6 (2002): 859–67.

Moore, Amelia. "Anthropocene Anthropology: Reconceptualizing Contemporary Global Change." *Journal of the Royal Anthropological Institute* 22, no. 1 (2016): 27–46.

Moore, Jason W. "The Capitalocene, Part I: On the Nature and Origins of Our Ecological Crisis." *Journal of Peasant Studies* 44, no. 3 (2017): 594–630.

Morrissette, Hannah K., Stacy K. Baez, Lisa Beers, Nadia Bood, Ninon D. Martinez, et al. "Belize Blue Carbon: Establishing a National Carbon Stock Estimate for Mangrove Ecosystems." *Science of the Total Environment* 870 (2023): 161829.

Mumby, Peter J. "The Impact of Exploiting Grazers (Scaridae) on the Dynamics of Caribbean Coral Reefs." *Ecological Applications* 16, no. 2 (2006): 747–69.

Mumby, Peter J., Alan Hastings, and Helen J. Edwards. "Thresholds and the Resilience of Caribbean Coral Reefs." *Nature* 450, no. 7166 (2007): 98–101.

Nelson, Sara Holiday. "Beyond The Limits to Growth: Ecology and the Neoliberal Counterrevolution." *Antipode* 47, no. 2 (2015): 461–80.

Nijbroek, Ravic. "Mangroves, Mudbanks and Seawalls: Political Ecology of Adaptation to Sea Level Rise in Suriname." PhD diss., University of South Florida, 2012.

Norgaard, Richard B. "Ecosystem Services: From Eye-Opening Metaphor to Complexity Blinder." *Ecological Economics* 69, no. 6 (2010): 1219–27.

Nuñez, Angel. "25 Years Ago in San Pedro-Garifunas in San Pedro," Ambergris Today, November 17, 2005. http://www.ambergristoday.com/content/25-years-ago/2005/november/17/garifunas-san-pedro?page=1.

Obermeyer, Nancy J. "GIS: The Maturation of a Profession." *Cartography and Geographic Information Science* 34, no. 2 (2007): 129–32.

Odum, Eugene Pleasants, and Gary W. Barrett. *Fundamentals of Ecology*. Thomson Brooks/Cole, 2005.

Oliver-Smith, Anthony, and Susanna M. Hoffman, eds. *The Angry Earth: Disaster in Anthropological Perspective*. Routledge, 1999.

Ong, Aihwa. "Neoliberalism as a Mobile Technology." *Transactions of the Institute of British Geographers* 32, no. 1 (2007): 3–8.

Onsrud, Harlan, and Werner Kuhn. *Advancing Geographic Information Science: The Past and Next Twenty Years*. Lulu.com, 2016.

Oreskes, Naomi, and John Krige. *Science and Technology in the Global Cold War*. MIT Press, 2014.

Oreskes, Naomi, Kristin Shrader-Frechette, and Kenneth Belitz. "Verification, Validation, and Confirmation of Numerical Models in the Earth Sciences." *Science* 263, no. 5147 (1994): 641–46.

Oxford Urbanists. "Vernacular Architecture and Climate Change," January 10, 2020. https://www.oxfordurbanists.com/oxford-urbanists-monthly/2020/1/5/vernacular-architecture-and-climate-change.

Paredis, Erik, Gert Goeminne, Wouter Vanhove, Jesse Lambrecht, and Frank Maes. *The Concept of Ecological Debt: Its Meaning and Applicability in International Policy*. Academia, 2009.

Patel, Hrishikesh, Sucharita Gopal, Les Kaufman, Matt Carleton, Chris Holden, et al. "MIDAS: A Spatial Decision Support System for Monitoring Marine Management Areas." *International Regional Science Review* 34, no. 2 (2011): 191–214.

Pearce, Jenny. "Natural and Unnatural Disasters in Central America." In *South America, Central America and the Caribbean 2003*, ed. Jacqueline West. Taylor & Francis, 2002.

Peck, Jamie. "Liberating the City: Between New York and New Orleans." *Urban Geography* 27, no. 8 (2006): 681–713.

Peck, Jamie, Nik Theodore, and Neil Brenner. "Neoliberal Urbanism: Models, Moments, Mutations." *SAIS Review of International Affairs* 29, no. 1 (2009): 49–66.

——. "Postneoliberalism and Its Malcontents." *Antipode* 41 (2009): 94–116.

Polanyi, Karl. *The Great Transformation: The Political and Economic Origins of Our Time*. Beacon, 1944.

Pott, Roberto. "Evaluating Local Access to Seafood in Urban Belize City." Master's diss., Duke University, 2010.

Povinelli, Elizabeth A. *Between Gaia and Ground: Four Axioms of Existence and the Ancestral Catastrophe of Late Liberalism*. Duke University Press Books, 2021.

Poynter, Jane. *The Human Experiment: Two Years and Twenty Minutes Inside Biosphere 2*. Basic Books, 2006.

Raffles, Hugh. *In Amazonia: A Natural History*. Princeton University Press, 2002.

Rahder, Micha. *An Ecology of Knowledges: Fear, Love, and Technoscience in Guatemalan Forest Conservation*. Duke University Press, 2020.

Ramos, Adale. "Climate Change Costing Central America US$105 Billion and Counting. . . " *Amandala Newspaper*, November 24, 2009. http://amandala.com.bz/news/climate-change-costing-central-america-us105-billion-and-counting/.

Reardon-Smith, Mardi. "In the Weeds: Thinking Through Anthropological and Other Social Science Approaches to Invasive Plant Species and Their Control." *Environment and Planning E: Nature and Space* 8, no. 2 (2025): 660–74.

Reider, Rebecca. *Dreaming the Biosphere: The Theater of All Possibilities*. University of New Mexico Press, 2009.

Ricker, B. A., P. R. Rickles, G. A. Fagg, and M. E. Haklay. "Tool, Toolmaker, and Scientist: Case Study Experiences Using GIS in Interdisciplinary Research." *Cartography and Geographic Information Science* 47, no. 4 (2020): 350–66.

Robertson, Morgan M. "The Nature That Capital Can See: Science, State, and Market in the Commodification of Ecosystem Services." *Environment and Planning D: Society and Space* 24, no. 3 (2006): 367–87.

Rose, Deborah Bird. *Shimmer: Flying Fox Exuberance in Worlds of Peril*. Edinburgh University Press, 2022.

Rosendo, Lucía Ruiz. "English and Spanish Medical Languages: A Comparative Study from Translation Point of View." *TRANS: Revista de Traductología*, no. 12 (2008): 231–46.

Roy, Ananya. *Poverty Capital: Microfinance and the Making of Development*. Routledge, 2010.

Rudolph, John L. *Scientists in the Classroom: The Cold War Reconstruction of American Science Education*. Palgrave Macmillan, 2002.

Runk, J. Velásquez, Gervacio Ortíz Negría, Leonardo Peña Conquista, Gelo Mejía Peña, Frecier Peña Cheucarama, and Yani Cheucarama Chiripua. "Landscapes, Legibility, and Conservation Planning: Multiple Representations of Forest Use in Panama." *Conservation Letters* 3, no. 3 (2010): 167–76.

Rutherford, Danilyn. "Affect Theory and the Empirical." *Annual Review of Anthropology* 45, no. 1 (2016): 285–300.

Sagan, Dorion, and Lynn Margulis. *Biospheres from Earth to Space*. Enslow, 1989.

Sahlins, Marshall. *Culture and Practical Reason*. University of Chicago Press, 2013.

——. *Islands of History*. University of Chicago Press, 1985.

Sanchez, Jose. "How Will San Pedro Get Rid of Undesirables on the Island?," March 19, 2010. https://edition.channel5belize.com/archives/29866.

Sawyer, Suzana. *Crude Chronicles: Indigenous Politics, Multinational Oil, and Neoliberalism in Ecuador*. Duke University Press, 2004.

Sayre, Nathan F. "The Politics of the Anthropogenic*." *Annual Review of Anthropology* 41 (2012): 57–70.

Schuhmann, Peter W. "Valuation of Ecosystem Services as a Basis for Investment in Blue Economies." In *The Caribbean Blue Economy*. Routledge, 2023.

Schumacher, Ernst Friedrich. *Small Is Beautiful: Economics as If People Mattered.* Harper & Row, 1973.

Shange, Savannah. *Progressive Dystopia: Abolition, Antiblackness, and Schooling in San Francisco.* Illus. ed. Duke University Press, 2019.

Shapiro-Garza, Elizabeth. "Contesting the Market-Based Nature of Mexico's National Payments for Ecosystem Services Programs: Four Sites of Articulation and Hybridization." *Geoforum* 46 (2013): 5–15.

Sheller, Mimi. *Consuming the Caribbean: From Arawaks to Zombies.* Routledge, 2003.

Shoman, Assad. *Belize's Independence and Decolonization in Latin America: Guatemala, Britain, and the UN.* Palgrave Macmillan, 2010.

Shotwell, Alexis. *Against Purity: Living Ethically in Compromised Times.* University of Minnesota Press, 2016.

Singh, Neera M. "The Affective Labor of Growing Forests and the Becoming of Environmental Subjects: Rethinking Environmentality in Odisha, India." *Geoforum* 47 (2013): 189–98.

——. "Introduction: Affective Ecologies and Conservation." *Conservation and Society* 16, no. 1 (2018): 1–7.

Smetacek, Victor, and Adriana Zingone. "Green and Golden Seaweed Tides on the Rise." *Nature* 504 (2013): 84–88.

Smith, Michael Garfield. *The Plural Society in the British West Indies.* University of California Press, 1965.

Solnit, Rebecca. *A Field Guide to Getting Lost.* Reprint ed. Penguin, 2006.

Srinivasan, U. Thara, Susan P. Carey, Eric Hallstein, Paul A. T. Higgins, Amber C. Kerr, Laura E. Koteen, Adam B. Smith, Reg Watson, John Harte, and Richard B. Norgaard. "The Debt of Nations and the Distribution of Ecological Impacts from Human Activities." *Proceedings of the National Academy of Sciences* 105, no. 5 (2008): 1768–73.

*Stanford News Service.* "Nature Subsidizes Humanity with Services Worth Trillions, Ecologist Says." February 6, 1997. http://news.stanford.edu/pr/97/970210daily.html.

Stanford Woods Institute for the Environment-Centers and Programs. "The Natural Capital Project," 2015. https://woods.stanford.edu/research/centers-programs/natural-capital-project.

Stehman, Stephen V. "Selecting and Interpreting Measures of Thematic Classification Accuracy." *Remote Sensing of Environment* 62, no. 1 (1997): 77–89.

Steinberg, Theodore. *Acts of God: The Unnatural History of Natural Disaster in America*. Oxford University Press, 2006.

Stewart, Kathleen. *Ordinary Affects*. Duke University Press, 2007.

Stocking, George W. "From Physics to Ethnology: Franz Boas' Arctic Expedition as a Problem in the Historiography of the Behavioral Sciences." *Journal of the History of the Behavioral Sciences* 1, no. 1 (1965): 53–66.

Stoler, Ann Laura. *Carnal Knowledge and Imperial Power: Race and the Intimate in Colonial Rule*. University of California Press, 2010.

Stone, Michael Cutler. *Caribbean Nation, Central American State: Ethnicity, Race, and National Formation in Belize, 1798–1990*. University of Texas, 1994.

Sullivan, Sian. "Banking Nature? The Spectacular Financialisation of Environmental Conservation." *Antipode* 45, no. 1 (2013): 198–217.

——. "Green Capitalism, and the Cultural Poverty of Constructing Nature as Service Provider." *Radical Anthropology* 3: 18–27.

——. "What's Ontology Got to Do with It? On Nature and Knowledge in a Political Ecology of the 'Green Economy.'" *Journal of Political Ecology* 24, no. 1 (2017): 217–42.

Sutherland, Anne. *The Making of Belize: Globalization in the Margins*. Bergin & Garvey, 1998.

Tansley, A. G. "The Use and Abuse of Vegetational Concepts and Terms." *Ecology* 16, no. 3 (1935): 284–307.

Taussig, Michael. *What Color Is the Sacred?* University of Chicago Press, 2009.

Thompson, Krista A. *An Eye for the Tropics: Tourism, Photography, and Framing the Caribbean Picturesque*. Duke University Press, 2007.

Thomson, Iain. "Heidegger's Aesthetics." In *The Stanford Encyclopedia of Philosophy*, ed. Edward N. Zalta, Fall 2015. http://plato.stanford.edu/archives/fall2015/entries/heidegger-aesthetics/.

Todd, Zoe. "An Indigenous Feminist's Take on the Ontological Turn: 'Ontology' Is Just Another Word for Colonialism." *Journal of Historical Sociology* 29, no. 1 (2016): 4–22.

Tsing, Anna. "The Buck, the Bull, and the Dream of the Stag: Some Unexpected Weeds of the Anthropocene." *Suomen Antropologi: Journal of the Finnish Anthropological Society* 42, no. 1 (2017): 3–21.

Tsing, Anna Lowenhaupt. *The Mushroom at the End of the World: On the Possibility of Life in Capitalist Ruins*. Princeton University Press, 2015.

Tsing, Anna, J. Peter Brosius, and Charles Zerner. "Introduction." In *Communities and Conservation*. Alta Mira Press, 2005.

Tsing, Anna Lowenhaupt, Andrew S. Mathews, and Nils Bubandt. "Patchy Anthropocene: Landscape Structure, Multispecies History, and the Retooling of Anthropology: An Introduction to Supplement 20." *Current Anthropology* 60, no. S20 (2019): S186–97.

Turner, Fred. *From Counterculture to Cyberculture: Stewart Brand, the Whole Earth Network, and the Rise of Digital Utopianism*. University of Chicago Press, 2008.

US Department of Commerce, National Oceanic and Atmospheric Administration. "What Is Sargassum?" Accessed July 29, 2019. https://oceanexplorer.noaa.gov/facts/sargassum.html.

Vaughn, Sarah E. "Disappearing Mangroves: The Epistemic Politics of Climate Adaptation in Guyana." *Cultural Anthropology* 32, no. 2 (2017): 242–68.

——. *Engineering Vulnerability: In Pursuit of Climate Adaptation*. Duke University Press, 2022.

——. "Erosion by Design: Rethinking Innovation, Sea Defense, and Credibility in Guyana." *Comparative Studies in Society and History* 64, no. 4 (2022): 849–77.

Velicu, Irina, and Gustavo García-López. "Thinking the Commons Through Ostrom and Butler: Boundedness and Vulnerability." *Theory, Culture & Society* 35, no. 6 (2018): 55–73.

Vellinga, Marcel. "The Inventiveness of Tradition: Vernacular Architecture and the Future." *Perspectives in Vernacular Architecture* 13, no. 2 (2006): 115–28.

Wainwright, Joel. "'The First Duties of Persons Living in a Civilized Community': The Maya, the Church, and the Colonial State in Southern Belize." *Journal of Historical Geography* 35, no. 3 (2009): 428–50.

Walker, Cameron. "Ecosystem Marketplace—Conservation Central's The Daily Show." Ecosystem Marketplace, 2006. http://www.ecosystemmarketplace.com/pages/dynamic/article.page.php?page_id=4474&section=news_arti.

Wang, Mengqiu, Chuanmin Hu, Brian B. Barnes, Gary Mitchum, Brian Lapointe, and Joseph P. Montoya. "The Great Atlantic Sargassum Belt." *Science* 365, no. 6448 (2019): 83–87.

Watts, Christopher. "Relational Archaeologies: Roots and Routes." In *Relational Archaeologies*. Routledge, 2013.

West, Paige. *Conservation Is Our Government Now: The Politics of Ecology in Papua New Guinea*. Duke University Press, 2006.

——. *Dispossession and the Environment: Rhetoric and Inequality in Papua New Guinea*. Columbia University Press, 2016.

Wiener, Norbert. *Cybernetics: Or, Control and Communications in the Animal and the Machine*. Wiley, 1948.

Williams, Eric. *Capitalism and Slavery*. Unknown ed. University of North Carolina Press, 1994.

Woods, Richard L., S. Tomlinson-Reid, and Alastair M. Reid. *The Field Guide to Ambergris Caye, Belize, C. A., Including Other Atolls*. Sue Tomlinson Guidebooks, 1988.

Worster, Donald. *Nature's Economy: A History of Ecological Ideas*. Cambridge University Press, 1994.

——. *Nature's Economy: The Roots of Ecology*. Sierra Club, 1977.

Yusoff, Kathryn. "Geologic Realism: On the Beach of Geologic Time." *Social Text* 37, no. 1 (138) (2019): 1–26.

——. "Politics of the Anthropocene: Formation of the Commons as a Geologic Process." *Antipode* 50, no. 1 (2017): 255–76.

Zimmermann, Francis. *Jungle and the Aroma of Meats: An Ecological Theme in Hindu Medicine*. Motilal Banarsidass, 2011.

# INDEX

Page numbers in *italics* refer to illustrations.

Printed and bound by CPI Group (UK) Ltd, Croydon, CR0 4YY

16/09/2026

14963454-0002